FOOD WEBS

FOOD WEBS

Stuart L. Pimm

The University of Chicago Press
Chicago & London

STUART L. PIMM is the Dorris Duke Chair of Conservation Ecology in the Nicholas School of the Environment and Earth Sciences at Duke University. He is the author of *The World According to Pimm: A Scientist Audits the Earth* and *The Balance of Nature? Ecological Issues in the Conservation of Species and Communities,* the latter published by The University of Chicago Press.

The University of Chicago Press, Chicago 60637
The University of Chicago Press, Ltd., London

Printed in the United States of America

11 10 09 08 07 06 05 04 03 02 1 2 3 4 5

ISBN: 0-226-66832-0 (paper)

Library of Congress Cataloging-in-Publication Data

Pimm, Stuart L. (Stuart Leonard)
Food webs / Stuart L. Pimm.
p. cm.
Originally published: London : New York : Chapman and Hall, 1982.
Includes bibliographical references and index.
1. Food chains (Ecology) 2. Biotic communities. I. Title.

QH541 .P56 2002
2002022736

⊚ The paper used in this publication meets the minimum requirements of the American National Standard for Information Sciences—Permanence of Paper for Printed Library Materials, ANSI Z39.48-1992.

Contents

Preface

Often the meanings of words are changed subtly for interesting reasons. The implication of the word 'community' has changed from including all the organisms in an area to only those species at a particular trophic level (and often a taxonomically restricted group), for example, 'bird-community'. If this observation is correct, its probable cause is the dramatic growth in our knowledge of the ecological patterns along trophic levels (I call these horizontal patterns) and the processes that generate them. This book deals with vertical patterns – those across trophic levels – and tries to compensate for their relative neglect. In cataloging a dozen vertical patterns I hope to convince the reader that species interactions across trophic levels are as patterned as those along trophic levels and demand explanations equally forcefully. But this is not the only objective. A limited number of processes shape the patterns of species interaction; to demonstrate their existence is an essential step in understanding why ecosystems are the way they are.

To achieve these aims I must resort to both mathematical techniques to develop theories and statistical techniques to decide between rival hypotheses. The level of mathematics is likely to offend nearly everyone. Some will find any mathematics too much, while others will consider the material to be old, familiar ground and probably explained with a poor regard for rigour and generality. However, a British student with 'A' level mathematics or his American counterpart with two semesters of college calculus will find nothing beyond his training and much that will be a revision of familiar ideas. It is for these students – who need to understand the biological assumptions in mathematical ecology and for whom texts on differential equations and linear algebra are too daunting – that this book is designed.

Developing theory using mathematical techniques is only one aspect of this book. The research programme that we call the field of population dynamics has not only clarified our ideas (e.g. those on the stability–complexity question) but has synthesized much existing knowledge. Most importantly, the theory has suggested new phenomena, including ones that we can observe or demonstrate experimentally in the field. The importance of theory to the field ecologist is clear, but what does the field ecologist offer to the theoretician? I shall argue that the possibilities available to the theoretician are so many that only by considering the field can he prevent his time being wasted in extensive investigations of theoretically possible yet biologically bizarre phenomena. My intent has been to place this book firmly on the interface between theory and observation.

In developing my ideas I owe a particular debt to John Lawton and Michael Rosenzweig. Much of the original work is based on joint papers with John, many of the ideas are his, and very few, if any, of the ideas have not been hotly debated over ale under the shadow of York Minster or while consuming tamales in the mountains of New Mexico. Only John's prior writing commitments prevented him from co-authoring this book, and it is poorer as a consequence.

Michael Rosenzweig, in addition to editing the book, has been a particularly close colleague. He has markedly shaped my ideas on ecology and provided encouragement without which I would have produced many fewer papers, no books, and certainly no grant proposals.

I am also in debt to those who reviewed all or part of this book. Michael Rosenzweig and John Coulson reviewed the entire document. Detailed chapter reviews were undertaken by W. Atmar, T. S. Bellows, J. E. Cohen, D. L. DeAngelis, M. P. Hassell, R. L. Kitching, J. H. Lawton, R. T. Paine, E. R. Pianka, P. W. Price, W. M. Post, R. D. Powell, R. J. Raitt, S. C. Stearns, N. C. Stenseth, and M. Williamson. Students in my classes (particularly, A. W. King, M. P. Moulton and M. E. T. Scioli) also provided useful comments. Errors no doubt remain, but the book would have been much the poorer without their advice. Much of the original work in this book was carried out under the tenure of grants from the National Science Foundation to Texas Tech University and from grants from the National Science Foundation and Department of Energy to Oak Ridge National Laboratory. My colleagues and the staff at Oak Ridge have been particularly supportive of my research on food webs.

I dedicate this book to my wife, June, and my parents, Hannah and Leonard.

Portal, Arizona

Acknowledgements

I gratefully acknowledge the authors and publishers of the original material that has been reproduced or modified to give the figures and tables in this book for their permissions. Full citations are given in the figure and table legends and at the appropriate places in the text together with the relevant entry in the bibliography. Figures 3.3, 3.4, 3.5, 4.9, 4.10, 4.11, 4.12, 4.13, 4.14, 4.15, 6A. 1, 8.1, 8.3, 8.4, and 8.5 and Tables 5.4, 6.1, 6A.1, 7.2, 7.3, 8.3, 8.4, 8.5, 8.6, and 8.7 are from papers produced while I was supported by the Oak Ridge National Laboratory (operated by the Union Carbide Corporation for the United States Department of Energy). Figure 4.3 is also reproduced with their permission. Figure 4.1 is modified from figures by R. T. Paine and is reproduced by permission from *The American Naturalist* (copyright 1966 The University of Chicago Press). Figure 4.4 is by R. M. May and reproduced by permission of Dr W. Junk, BV Publishers. Figure 6.3 is modified from Figure 2.6 in J. Phillipson's monograph in The Institute of Biology's *Studies in Biology* series published by Edward Arnold (Publishers) Ltd. Figure 6.5 is by permission of Blackwell Scientific Publications Ltd and includes material from a figure by T. R. E. Southwood. Figures 6.1 and 6.2 are based on figures by T. M. Zaret and R. T. Paine, Figure 7.1b on a figure by D. C. Force and Table 5.3 is a table by L. E. Hurd, M. V. Mellinger, L. L. Wolf and S. J. McNaughton and are reproduced from *Science* (copyrights 1973, 1960 and 1971 respectively, by the American Association for the Advancement of Science). Figure 6.10 is from a paper by R. L. Kitching in a volume published by Plexus Publishing, Inc. Figure 6.8 is modified from a paper by S. L. Pimm and J. H. Lawton in *Nature* (Volume 268, page 329), Figure 7.2 and Table 7.1 are from a similarly authored paper in *Nature* (Volume 275 page 542), and Figure 4.2 is modified from a paper by M. R. Gardner and W. R. Ashby in *Nature* (Volume 228, page 784) copyright 1977, 1978 and 1970 respectively by MacMillan Journals Ltd. Tables 5.1 and 5.2 are from a paper by S. J. McNaughton and are reproduced with permission from *The American Naturalist* (copyright 1977, The University of Chicago Press). Tables 9.1 and 9.4 are excerpted from Tables 5 and 9 in *Food Webs and Niche Space* by J. E. Cohen (copyright 1978 by Princeton University Press).

In addition I thank the editors of these journals for permission to reproduce, modify and excerpt figures and tables: *Ecology* (Figures 6A.1, 7.1a, Tables 1.1, 6.1, 6A.1, 7.2, 7.3), *Oikos* (Figures 3.3, 4.9–4.15, Table 5.4), *The Journal of Animal Ecology* (Figures 1.2b, c, 6.9, 8.1, 8.3, 8.4, 8.5, Tables 8.3, 8.4, 8.5, 8.6, 8.7) and *Theoretical Population Biology* (Figures 4.7, Table 8.1).

Conventions and definitions

X_i	The *i*th species *or* its density.
$\dot{X}_i$	The rate of change of X_i, that is, dX_i/dt.
X_i^*	The non-trivial equilibrium density of X_i, that is, where $\dot{X}_i = 0$ and $X_i \neq 0$.
a_{ij}	The interaction coefficient between X_i and X_j; the effect of one individual of X_j upon the growth rate of an individual of X_i.
b_i	The *per capita* rate of increase or decrease of X_i in the absence of any other species.
c_{ij}	Elements of the Jacobian matrix.
x_i	$X_i - X_i^*$.
$f(x), g(x)$	Functions of the variable(s) within parentheses.
λ_i, λ_{max}	Eigenvalues, the maximum eigenvalue. For their calculation see Chapter 2.
$\sum_{i=1}^{n} X_i$	$X_1 + X_2 + \ldots + X_n$.
$\exp(a)$	e^a, where e is the base of natural logarithms.
$\ln(a)$	*a* expressed as a natural logarithm.
$X_{1,t}$	The density of X_1 at time *t*.
$\partial y/\partial x_i$	The partial derivative of *y* with respect to x_i; obtained by differentiating *y* with respect to x_i and keeping all the other possible variables ($x_j, j \neq i$) constant.
n	The number of species in a system; sometimes the number of species of prey in a system.
m	Usually, the number of species of predators in a system.
C	See connectance.
P	A proportion (usually obtained by Monte-Carlo simulations and involving the proportion of random webs that exceed an observed web in some character).
α	The probability of a hypothesis given an observed result.
autotroph	Species which can obtain energy directly from sunlight and/or chemical sources and which do not require other organisms as food.
basal species	Species which feed on no other species in a food web.
connectance	The proportion of the possible interspecies interactions that are nonzero.
determinant	See Chapter 2, page 32.
detritus	Dead animal and plant matter.

detritivore	Animals or plants that feed on detritus.
ectotherm	Animals which do not usually maintain a constant body temperature.
endotherm	Animals which usually maintain a constant body temperature.
feasible	A model where all $X_i^* > 0$.
heterotroph	Animals which must feed on other organisms to survive.
Jacobian matrix	See Chapter 2, Appendix 2C.
loop	A pattern of species interaction where species A feeds on species B which feeds on species A, or A feeds on B, B on C, C on A, etc.
monophage	A species which feeds on only one other species.
oligophage	A species which feeds on few other species.
omnivore	A species which feeds on more than one trophic level.
phytophage	A species which feeds on plants.
polyphage	A species which feeds on many different species.
predator	A species (animal or plant) which feeds on other species.
prey	A species (animal or plant) on which other species feed.
return time	See Equation (2.42). The time it takes perturbations to reach 1/e (about 37%) of their initial value in a stable system.
saprophage	A species which feeds on decaying plant material.
singular	A system whose determinant is zero; see Chapter 2, page 32.
top-predator	A species on which nothing else feeds.

Foreword

WHY REPRINT A TWENTY-YEAR-OLD SCIENCE BOOK?

Scientific knowledge accumulates so rapidly that papers just a few years old seem quaint in their assumptions and methods. *Food Webs*, published twenty years ago, is antediluvian by these standards, so why this reprint? One answer might be that it is a classic—something to be appreciated by those who study history and the origin of ideas—and that the lessons it teaches may yield insights into present science. To those who read this reprint with history in mind I wish only the best. It is not my reason for this reprint's existence.

In the twenty years since this book was published, we have added two billion humans to the planet, cleared about three million square kilometers of tropical forests, over-harvested a large number of fisheries, caused the warmest years in recorded history and witnessed countless other human actions have massive environmental impact. This is not the place to review these, nor justify my optimism that we can do what is required to protect our planet for countless future generations. I do so elsewhere (Pimm, 2001). Nor is this the place to outline the agenda of actions for that protection. In developing that agenda, a broad group of colleagues and I (Pimm et al., 2001) make an insistent and unanimous recommendation to train a new and much larger generation of professionals to tackle environmental problems. By "professionals" we mean those with skills far more wide ranging than just ecologists. Nonetheless, however much age teaches me that students must also speak law, politics, economics, and other languages foreign to scientists, ecology is a core skill. I still teach basic ecological concepts, including food webs, even as my students arrive from a class on international science policy and depart to another on economics. Importantly, many others are still teaching food web ideas too—something I know from those who photocopy chunks of this book and from the inquiries I have about reprinting it. Simply, at a time when our students learn even more skills and must do so with a keen sense of urgency, this book's material is a useful teaching resource. To protect Nature, we must have some understanding of her complexities, for which the food web is the basic description.

The justification of this reprint, then, is to provide, within this foreword and the original text, the materials to bring its reader up to speed with current ideas and controversies. Moreover, it must do so faster than simply reading the most current literature. There are some subjects where the current literature is far preferable to the old, which may be premature or conceptually muddy and confusing. Starting from the beginning—retelling a subject's history—need not be the only way. Yet,

I think it is for food webs. That's why *Food Webs* is being reprinted. It's also why this foreword has the structure it does.

Food Webs has four major themes:

(i) The majority of communities consist of stable populations, that is, those showing a tendency to return to an equilibrium density when perturbed from it.

(ii) The requirement of stability imposes constraints on the patterns of how species should be connected—that is, food web structure.

(iii) Empirically, food webs are structured—they differ from what one would expect by chance and do so in ways anticipated by the theory.

(iv) Food web structure affects community dynamics.

This book also introduces two broad methods—how to build multi-species models to investigate these topics and how to conduct field experiments to test them.

THE NATURE OF POPULATION DYNAMICS

To develop a dynamical theory of food webs, I needed to show that stable populations were the norm (pp. 8–11). Ecologists have debated the nature of population change for decades, arguing the relative merits of density dependence—which is necessary if densities are to have some central tendency; density independence—for which populations vary without bounds; and density vagueness—by which populations vary wildly but are constrained at rarely encountered lower and upper limits. Whatever the merits of these explanations, my key concern was their generality. This requires the comparative study of populations.

The late Jim Tanner had attempted such a study and my addition to his results broadly followed his recipe. Statistical difficulties notwithstanding, fitting the parameters of the familiar r/K model—r is the population's growth rate when well below its equilibrium level, K—is as simple an exercise as one could imagine. From Tanner's study and my analyses of British birds (pp. 10, 11), I concluded that the assumption of stability was a sensible one.

In the intervening two decades there have been three important advances.

The Global Population Dynamics Database (GPDD). Accumulating population time series is a lengthy process, requiring ecologists to devote a lifetime to the same species at the same location. (Indeed, some of the longest series span several scientific generations.) The work is not only tedious but is notorious for not being financially rewarding. Not surprisingly then, there seemed to be few long-term studies. Yet as I searched the literature in the years following the publication of *Food Webs*, I realized there were far more than I had expected. My enlarged collection formed a resource for my next book *The Balance of Nature? Ecological Issues in the Conservation of Species and Communities* (Pimm, 1991). John Lawton, of the Centre for Population Biology, Imperial College at Ascot, had also been compiling time series. We met in Tennessee in June 1994 and agreed to a joint effort to search the literature and make available as many series as we could to the ecological community. The National Center for Ecological Analysis and Synthesis (University of California, Santa Barbara) soon joined that effort.

The NERC Centre for Population Biology built the GPDD, which now consists of more than 4,500 time series of population abundance, each longer than ten years. It encompasses over 1,800 animal species across many geographical locations. The GPDD is updated continuously with new information from published and unpublished sources. It is freely available and fully searchable: http://cpbnts1.bio.ic.ac.uk/gpdd/.

Statistical modeling. The techniques for modeling populations have improved spectacularly and now allow ecologists, inter alia, to dissect the time lags that lead to complex cycles, the interactions with other species, and the long-term impacts of climate events. Bjørnstad & Grenfell (2001) provide an excellent review of this progress. We now know that the complex bestiary of possible population dynamics anticipated by Fig. 1.2 is realized, plus there are many more possibilities than any of us dreamt of.

The nature of population change revisited. I found the idea of a species' equilibrium density embodied in the r/K equation to pose enormous problems when viewed in the context of the food web. This thing called K—the equilibrium density—is the integration of all the other species present in the community—predators, prey, competitors, mutualists, and diseases. Are all of these species expected to stand still politely, while our species of interest returns to its equilibrium? The food web view demands that we think of population change in a multi-species context. Change one species and, in time, all the others will change too. (That idea is at the heart of the explanation of why food chains are short, but I am getting ahead of the story.)

The idea that species depend on other species, which in turn depend on others, leads to the idea of dynamical effects imposed upon other dynamical effects, imposed upon yet others, and so on, throughout a complex of interactions that the food web describes. It suggests a model of dynamical change approximating "red noise"—where small, short-term changes are imposed on larger, longer-term changes, which are imposed on yet larger, longer changes, and so on. With that view, populations may appear to show some equilibrium in the short term, but that level will change over the longer term—and change more, the longer one looks at the population.

That idea prompted an analysis (Pimm & Redfearn, 1988) that showed that population variance increased over time—the "more time means more variation" as John Lawton put in his News and Views that accompanied the article in *Nature* (Lawton, 1988). John Halley told me that the analysis and particularly the explanation of it in the preceding paragraph "greatly angered" him when I presented the work at the CPB. He and Pablo Inchausti, armed with the GPDD, investigated the idea (Inchausti & Halley, 2001).

For the analysis, they used all annual series longer than 30 years. The GPDD contains 544 such series, representing 123 species. Their results confirm and greatly extend my work (Pimm & Redfearn, 1988, Ariño & Pimm, 1995) and that of others that population variability increases with time series length. In over 95% of their series, there is an increase in population variability, but it decelerates with time series length. This deceleration need not imply convergence to an upper limit. For the majority of ecological series, variance fails to converge to any limit, at least over the time scales the data encompass.

Traditional models of density-dependent growth imply the existence of an equilibrium that confines the population abundances to a range of values about equilibrium. For such populations, the variance *should* converge to a clear limit. By contrast, density-independent dynamics, subject to the vagaries of environmental noise, show a random walk over time. For such dynamics, the variance grows linearly with time. Inchausti & Halley (2001) conclude—as did Arturo Ariño and I with our far fewer series—that the dynamics of animal populations typically lie somewhere between the two extremes.

It is possible to adopt a worldview of all populations undergoing a random walk. Steve Hubbell, in *The Unified Neutral Theory of Biodiversity and Biogeography,* has done just that (Hubbell, 2001). His predictions are numerous and compelling and, perhaps given the number of Inchausti and Halley's populations that appear close to (or indistinguishable from) random walks, we should not be surprised. Equally, the original assumption required to develop a theory of food webs survives intact, for most species dynamics are more bounded than random walks. A sensible assumption for community dynamics is of a multi-species equilibrium about which species are constantly attracted but which undergo a complex dance as environmental noise and a myriad of interspecific interactions drag them away from it.

BUILDING AND ANALYZING MULTI-SPECIES MODELS

Chapters 2, 3, and 4, (pp. 12–83), deal with mathematics that a friend in comments about *Food Webs* called "both dull and daunting". He, like the rest of us, has to use them anyway and I presented them at some length because I found the alternative explanations just horrible. The analysis of systems of differential equations typically came toward the end of introductory textbooks on the subject and required substantial preparation in calculus. I found that undergraduate courses sometimes did not cover the topic at all in a first course. That meant that students had to sit through a couple of courses on calculus, then one on differential equations, and only then get into one that helped. I needed something to teach that was much more direct. Understanding how multi-species models behave and how best to characterize them is an essential skill in ecology. Two decades later, I'm still using these chapters and seeing them photocopied more than others. Yes, they require basic calculus and some algebra.

With those skills, it is then a matter of developing the intuition about how complex systems behave. The advent of personal computers has made modeling much easier, and there are excellent simulation packages. More importantly, one can do so much using ubiquitous spreadsheet packages and, in doing so, check the intuitions the mathematics of these chapters provide.

For example, Chapter 6 asks: How quickly will an ecosystem recover when we subject it to transient shock? Suppose phytoplankton in a lake were shocked with a pulse of nutrients. The intuition is that the phytoplankton would first increase, then decrease as they use up the nutrients. Alone, they could probably recover normal levels quickly. If the lake also contained phytoplankton, zooplankton, and fish, the increased phytoplankton would cause the zooplankton to increase, and then the increased zooplankton would cause the number of fish to increase. Now, while the fish remain unusually abundant, their prey, the zooplankton will be rare. Consequently, their prey, the phytoplankton, will be unusually abundant.

Simply, no component of the system can return to equilibrium until all the others do so. To study only one component is to hear the plop of the stone in the pond, but not see the ripples spread. The recovery time is likely to depend on the length of the food chain. The longer the chain, the further those ripples have to travel. I lay out the requisite mathematics in Chapters 2, 3, 4, and 6, but the means to model the intuition is as close as your nearest spreadsheet.

The tinker toy models I used would first consider an equation for the phytoplankton limited by nutrients:

$$dX_1/dt = X_1(b_1 - a_{11}X_1). \tag{1}$$

The growth rate of the phytoplankton, dX_1/dt, depends on the size of its population, X_1, its intrinsic growth rate, b_1, and a limitation imposed by the shortage of nutrients, a_{11}. (This is the familiar "r and K" population model in a different guise. It is one that allows us to add other trophic levels more readily.) The population size approaches "K," its equilibrium value b_1/a_{11} from any value of X_1. The question is how fast it will approach that value.

Now let's add another trophic level, the zooplankton:

$$\begin{aligned} dX_1/dt &= X_1(b_1 - a_{11}X_1 - a_{12}X_2) \\ dX_2/dt &= X_2(-b_2 + a_{21}X_1). \end{aligned} \tag{2}$$

The phytoplankton now suffers predation from the zooplankton. The zooplankton die off if there are insufficient phytoplankton to support them, that is, when $(-b_2 + a_{21}X_1) < 0$ or $X_1 < b_2/a_{21})$.

We can keep on adding levels—the three trophic level model is

$$\begin{aligned} dX_1/dt &= X_1(b_1 - a_{11}X_1 - a_{12}X_2) \\ dX_2/dt &= X_2(-b_2 + a_{21}X_1 - a_{23}X_3) \\ dX_3/dt &= X_3(-b_3 + a_{32}X_2). \end{aligned} \tag{3}$$

One way to explore the equations' behavior is to simulate them. We can replace the differential equations with calculations using small but finite time steps, Δt.

The smaller the step, the closer these finite difference equations will approximate the differential equations (that is $\Delta X/\Delta t \approx dX/dt$ for small Δt). The idea is that

$$X_{t+\Delta t} = X_t + \Delta X. \tag{4}$$

The finite difference approximation for equation (1) would be:

$$\Delta X_1 = \Delta t \,.\, X_1(b_1 - a_{11}.X_1) \tag{5}$$

As an example, I have investigated the three species system

$$dX_1/dt = X_1(1.0 - 0.01X_1 - 0.1X_2)$$
$$dX_2/dt = X_2(-1.0 + 0.02X_1 - 0.1X_3)$$
$$dX_3/dt = X_3(-1.0 + 0.5X_2) \tag{6}$$

using a time step, $\Delta t = 0.1$, and initial values of the three species of 50, 10, and 3.

You can do this at home (or wherever you keep your computer). Put these first three numbers into row 1 of a spreadsheet, into columns A, B, and C respectively, thus:

50	10	3

Add three formulas into the three columns of the next row

Row 2, column A	=	A1+(0.1)*A1*[1−(0.01*A1)−(0.1*B1)]
Row 2, column B	=	B1+(0.1)*B1*[−1+(0.02*A1)−(0.1*C1)]
Row 2, column C	=	C1 +(0.1)*C1*[−1+(0.5*B1)],

and ask the computer to calculate the new values, which are

47.5	9.7	4.2.

Spreadsheets now have the convenient feature of allowing one to "fill down" the calculations as many rows as one wants. The next row will contain the formulas for A3, B3, and C3, and calculate them as

45.386	9.244	5.817

and so on for 500 rows.

To simulate just the phytoplankton and the herbivore, set the first value of C to 0, and it will stay there. To simulate just the phytoplankton, set B = C = 0.

So, armed with only a spreadsheet, one can explore the section on the dynamical constrains of food chain length—and, for that matter, most of the other theories in the book. (The one warning is that the crude assumption of making $\Delta t = 0.1$ will fail for some models and a smaller value will be necessary.)

Do not bother reading the stuff on species deletion stability on pages 47–49 and 77–82—at least not just yet. It is not that it is wrong. My reasons for writing it were that I wanted to know the consequences of really bashing communities—taking out entire species—rather than just tweaking the densities of some populations. The ideas this generated kept me busy for another decade. At issue is how *resistant* communities are to change. These were the topics of my next book (Pimm, 1991) and are something to which I shall return at the end of this foreword.

HOW TO PARAMETERIZE MULTI-SPECIES MODELS

When *Food Webs* was first published, the best I could do was to guess parameter values and to assign values randomly over ecologically reasonable intervals. Figure 1.1 (p. 5) laid out the book's main argument: real communities are composed of species with dynamically stable populations. We can build food web models with different structures and with their parameters selected randomly over ecologically plausible intervals. We predict that those structures that rarely yield stable systems will not be common in nature.

Those with a more intimate knowledge of particular communities could go much further—and parameterize known food web structures with informed estimates of the parameters involved. A benchmark paper was de Ruiter et al. (1995). In discussing their work, Lisa Manne and I (1996) invoked the imagery of Rube Golberg.

Recall a typical Rube Goldberg contraption with a long, complex, and vulnerable chain of processes to achieve some simple end. We do not meet objects like this in the real world and they obviously would not work. When one examines food webs such as the one presented by de Ruiter, we ask the same questions. Shoud we see objects like it in Nature? Will it work, and, if so, then why? The analysis by de Ruiter et al. shows that this and six other soil-based food webs do likely "work." That is, these systems are likely to be dynamically stable. Unlike Goldberg contraptions that will fall apart, these food webs will not.

The structure of this web comes from the authors' knowledge of the system. But what about their interaction terms? These require several kinds of information. De Ruiter et al. took the biomass of a species to be the average annual population size of the species; call this X_i*. The feeding rate, F_{ij}, of a predator of density Xj on a prey of density X_i is modeled using Lotka-Volterra dynamics. This assumption leads us to equate the feeding rate F_{ij} with $c_{ij}X_i^*X_j^*$, where c is a constant particular to the two species involved. De Ruiter et al. estimated each feeding rate directly, then estimated the per unit effect of predator X_j on prey X_i as F_{ij}/X_j or $c_{ij}X_i^*$.

While the prey species loses F_{ij} to the predator per year, the predator's gain is much smaller. (When a rabbit is running for its life, the fox chasing it is merely running for its supper.) The predator's gain must be reduced by the fraction of the prey's tissues that it can assimilate (the assimilation efficiency) and the fraction of the assimilated tissue that it can convert into new biomass (the production efficiency). These efficiencies are reasonably well known for many groups of species.

With these estimates in hand, the authors asked two questions. Where are the fragile linkages within each web? Are these real food webs special compared to imaginary webs that we might create using different assumptions?

To tackle the first question, de Ruiter et al. calculated the impact of each pair interactions on the stability of the food web, by varying their magnitude and then calculating the probability that the matrix will become unstable. They allow each of the pair of interaction strengths to take a random value in the range zero to twice the estimated strength of each particular interaction. They analyze the stability of the matrix of interaction strengths using the methods outlined in Chapters 2 and 3.

The impacts on web stability were not obviously correlated with the biomass of the species involved. Nor did the magnitude of the interaction strengths obviously correlate with impacts of stability: some of the sensitive interactions involved strong interactions and other weaker ones. Rather, it is the food web "patterning" that is crucial to stability. The highly interconnected trophic interactions among the bacteriophagous nematodes, fungivorous nematodes, predatory nematodes, nematophagous mites, predatory collembola, and predatory mites were crucial in terms of preserving stability. This result confirms Chapter 4's general insight: it is the parts of the food web where the trophic connections are most complex that are important to its stability.

To address the second question, de Ruiter et al. compared the stability of four types of interaction matrices for each of the seven food webs, by doing 100 runs with different randomizations. The four types are "lifelike" matrices, using the estimated interaction strengths and observed patterns of trophic interaction (as already described); "disturbed" matrices, with the estimated interaction strengths, but where the patterns of trophic interactions were randomly permuted; and two different simulations where the observed trophic patterns were maintained but the interaction strengths were sampled randomly from different intervals.

The disturbed matrices were the poorest—they were less likely to be stable than any other alternatives. This result supports the conclusion that the patterns of trophic interactions are unusual in a statistical sense. The patterns we observe tend to be those consistent with stability: randomized patterns, even with the same interaction strengths, produce "contraptions"—systems much less likely to work.

The lifelike matrices were the "best buy"—they were the most likely to contain stable systems. This suggests that the parameter values are important too. The two sets of simulations with the same trophic patterns but different parameter values were less likely to be stable.

Together, the results point to a simple, but important, conclusion. Despite the inevitable uncertainties in producing the food web, the results are quite unusual. Certainly, some parts are more fragile than others. Perhaps they indicate the part of the system where we understand the dynamics the least. Overall, the structures have the parameter values and patterns of interaction that make them far more likely to work than we would expect by chance.

THE COMPARATIVE STRUCTURE OF FOOD WEBS

Chapter 5 is short and starts with a discussion of empirical results about stability. I want to postpone the discussion of the first five pages, for they are again the subject of the consequences of food web structure—a subject to which I shall return. Pages 89–91 discuss the relationship of connectance to species number.

Connectance and Linkage Density

The simplest question one can ask of a food web is how connected it is. Connectance is the fraction of possible inter-specific links that realized. The reason to use connectance was a theoretical one—it played a role in May's famous result that stable systems would be those with a sufficiently small connectance (1972). As Chapter 5 explains, connectance depends on the number of species, which I called **n**. Joel Cohen and his colleagues (1990), more sensibly, have concentrated on the relationship between the number of species and the total number of trophic links, L. His original claim was that *linkage density*, **d,** was likely to be constant,

$$L = \mathbf{d}n, \tag{7}$$

so that food webs were likely to "scale invariant" in this and other properties to be discussed presently. Were equation (1) to be the case, connectance would decline hyperbolically as Fig. 5.1 suggests.

By 1991, when Joel Cohen, John Lawton, and I joined to write a review about our combined efforts to elucidate food web structure (Pimm et al., 1991), we agreed that this was the pattern least likely to survive detailed scrutiny. Averaged over a much larger collection of food webs in the range from 3 to 48 species, the average number of linkages [E(L)] is roughly twice the number of species in any given web [i.e. $E(L) = 2S$; $\mathbf{d} = 2$]. The original description of this pattern noticed that a power-law $E(L) = kn^{1+\varepsilon}$, for some small positive ε, was also a viable description of the data, and that future data on webs with large numbers of species would have to distinguish the alternatives. With the few larger webs in hand by 1991, a power law with ε probably between 0.3 and 0.4 indeed seems reasonable. (Joel Cohen compiled the available food webs, has constantly undated the collection since, and they are available from him [Cohen, 1989a).

Gary Polis and Neo Martinez were both sharply critical of low estimates linkage densities. Williams & Martinez (2000) discuss a sample of species-rich webs with 25 to 92 species, for which the linkage density ranges from 2.2 to 10.8. The highest linkage densities come from Martinez (1991) and Polis (1991). They may be right, but one of the difficulties with food web studies has been the problem of where to stop drawing connections. Yes, I eat rice, beans, potatoes, chicken, and occasionally seaweed wrapped around sushi and durian fruits (when I can get them). At what stage should I stop drawing trophic connections in my own personal food web? Martinez and Polis may be simply listing more connections than other workers.

The best solution is where one investigator compares two or more webs from his own work. The benchmark here was the work by Karl Havens who compared the food web connectance of different lakes (Havens, 1992). On page 89, I had supposed that "each species in a community feeds on a number of species of prey that is independent of the total number of species in the community" and called this the most "parsimonious assumption." Havens both agreed and disagreed. Durians excepted, my diet doesn't greatly expand when I move into a species-rich tropical forest, I still eat my usual rice and beans. So the addition of many species to a food web will not alter my feeding preferences. Species that select particular species of prey will surely follow suit and linkage density will remain constant and not increase as the size of the food web increases. On the other hand, filter feeders in lakes, for example, select prey based on their size. The more species of prey there are, the more that will be of the right size. Linkage density will then increase in direct proportion to the number of species present. Havens separated the species in his lakes into those expected to select prey species and those that should be indiscriminant. He found what he expected. This suggests that food web linkage density will be somewhere between being constant and increasing in proportion to the number of species.

Other Features

Cohen's first book on food webs preceded mine by three years (Cohen, 1978) and in the interval between its publication and our joint 1991 review, he noticed other features that were generally conserved across food webs. (Cohen et al. [1990] compiles those papers into one volume and adds additional material by way of added explanation.)

(i) Trophic cycles occur when species A eats species B and B eats species A, or A eats B, B eats C, and C eats A and so on. Such cycles are generally very rare (Cohen, 1978, p. 186).

My marine ecologist friends howled in pain whenever I said this quote during my seminars during the mid-1980s. In marine systems, it is quite common for fish to eat their way up a food chain as they grow, starting as planktivores when small and ending up eating fish that eat fish that eat zooplankton that eat phytoplankton. This doesn't necessarily lead to trophic cycles. In some cases, a fish (call it species A) in the diet of this adult top-predator (species B) had a great liking to the eggs or early larval stages of the top-predator. In a special way A eats B and B eats A. So that I could show my face at marine meetings, I joined with Jake Rice to investigate the dynamics of this phenomenon (Pimm & Rice, 1987). Using the modeling methods of Chapters 2–4, we showed that eating one's way up a food chain did not destabilize it as much as feeding on all lower trophic levels simultaneously (omnivory, to be discussed below) and that some of these life history cycles were dynamically feasible. Polis (1991) found complex trophic cycles in scorpions, but beyond this, the topic has not received much attention.

> wiring diagram of a food web or the Internet or the metabolic network of the bacterium *Escherichia coli*? Are there any unifying principles underlying their topology? (2001)

He continues by noticing empirical studies of food webs, electrical power grids, cellular and metabolic networks, the World Wide Web, the neural network of the nematode *Caenorhabditis elegans*, the citation networks of scientists, and the 'old-boy' network—the overlapping boards of directors of the largest companies. The Internet now makes it possible to search these and many other networks in a way that we could not have twenty years ago. (I just love the fact that he always mentions food webs first.)

His particular interest is an extension of the notion of "six degrees of separation"—that your friends' (1) friends' (2) friends' (3) friends' (4) friends' (5) friends' (6) encompass all 6 billion of us on the planet. You only need to have twenty-six friends, each one of whom has to know twenty-five other friends, and so on. The catch is that those twenty-five other friends must not include your twenty-five friends. While one can design networks with this property (and Strogatz shows how), the usual way to connect everyone is for there to be some broad power law to the number of connections. Many nodes have few "friends," fewer

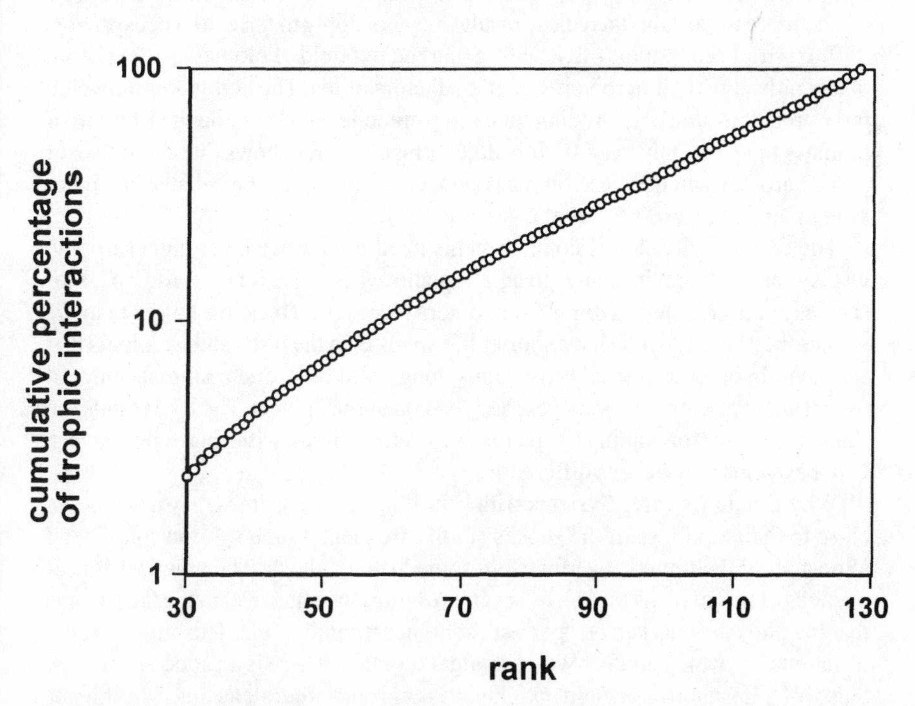

Figure F.4. The logarithm of the cumulative number of ~1500 trophic interactions against the rank for 128 species of predators in the rainforest food web of Reagan and Waide (1996). Rank is defined so the most trophically specialized species have rank 1 and the most trophically generalized species has rank 128.

retical food webs (Fig. F.1). (In doing all this, there are obvious parallels to the de Ruiter et al. study discussed earlier; food web ecologists can now estimate the parameters of their systems.)

The final step was to use these numbers to calculate both the resilience and rate of nutrient recycling in each year. In 1986, the recycling was slightly tighter than in 1984. The addition of the extra level increased the return time from an estimated twenty-eight days in 1984 to over 200 days in 1986. Adding the extra trophic level alters the distribution of the phosphorous. Freed from predation from planktivores, *Chaoborus* increases dramatically, and its prey, the herbivorous zooplankton, decrease. The major phosphorous stocks moved to higher trophic levels (piscivores, *Chaoborus*) in 1986, and these species have slower turnovers than the zooplankton and seston.

The predicted values are interesting for reasons other than their confirmation of theoretical predictions. There are at least two reasons for the increase in return time with trophic levels. First, the number of levels itself may matter. More levels must often slow the transit of nutrients through the system, even if the converse is possible. Second, species at higher trophic levels are larger and longer-lived in aquatic ecosystems. In this lake, shifting stocks from algae (which live days) to fish (which live years) slows the nutrient transfers. If this second mechanism is by far the more important, then these results may not apply to terrestrial ecosystems.

Terrestrial ecosystems often show a marked pyramid of biomasses. Terrestrial plants outweigh their herbivores often by factors of ten. The herbivores outweigh their predators similarly. Adding an extra trophic level moves the distribution of biomass upwards only very slightly. In contrast, Fig. F.1 shows that the stocks of phosphorous (which reflect biomass and vice versa) can be greater at higher rather than lower levels in aquatic systems.

The dynamically slower compartments need not always be at higher trophic levels—as they are in many pelagic aquatic webs. The relative life spans of species at different levels differ from system to system. Trees live longer than the insects they house, which have shorter life spans than the birds and mammals that eat them. In other terrestrial ecosystems, long-lived birds and mammals may eat insects that typically live for a year and feed on annual plants. The lake results are stacked in favor of finding an increase in return times with increasing levels. Other systems may behave differently.

Why should we care? Systems with very long recovery times may never come close to their equilibria if the shocks are too frequent. Some systems may spend almost all of the time recovering from some historic shock. The practical consequences of this may not always be severe. Adding a trophic level to the lake means that the phosphorous can be "parked" at higher trophic levels. This improves the water quality, which equates with low algal biomass. There is an added benefit. A pulse of phosphorous would quickly pass through the algae into the higher trophic levels. There it receives only a temporary parking permit—but it is a permit with a 200-day return time.

The section in this book on the dynamics of food chain lengths includes a

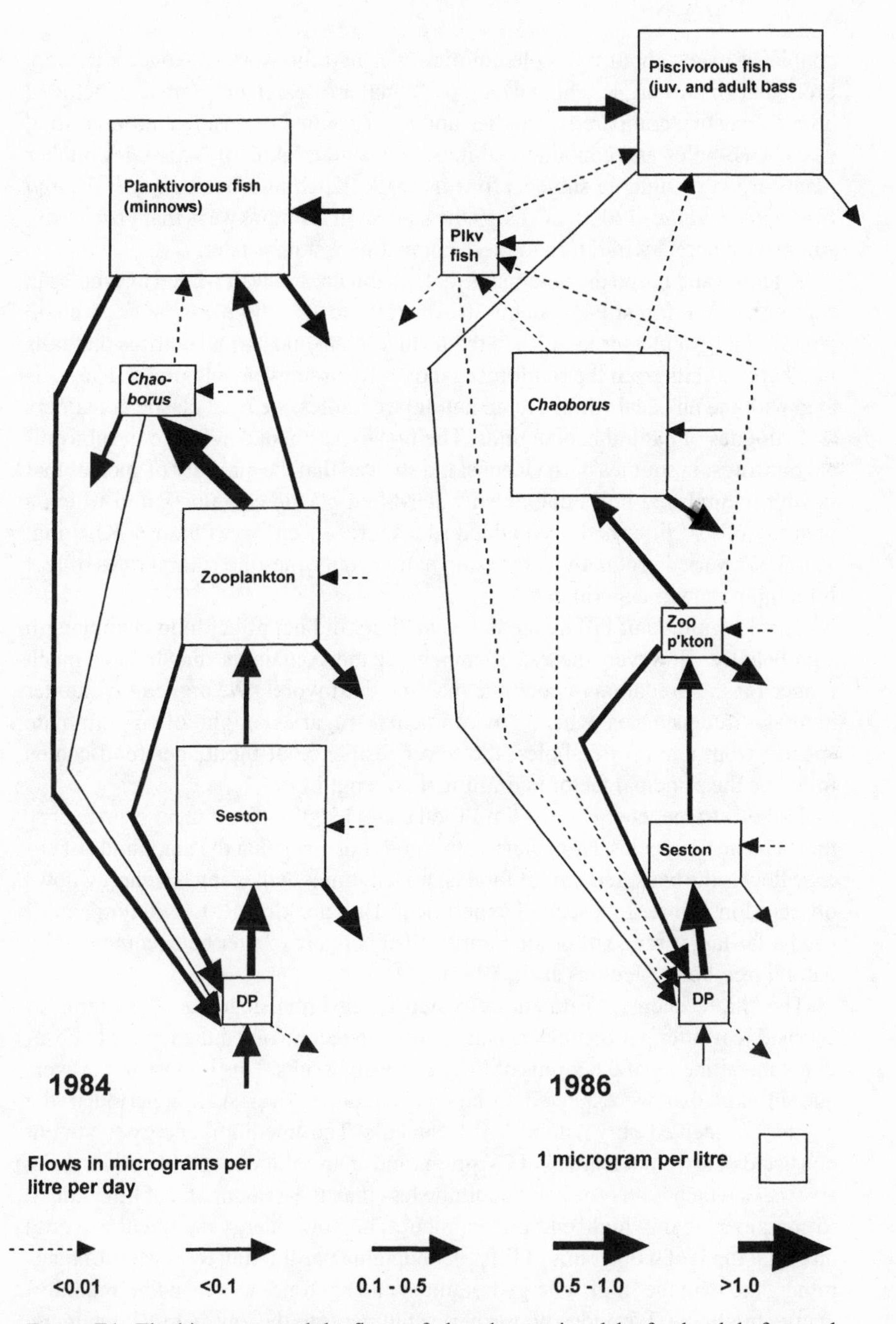

Figure F.1. The biomasses and the flows of phosphorous in a lake food web before and after an experimental addition of the top-trophic level.

couple of pages about tree-hole communities and the work of Roger Kitching. From that small start, Kitching developed a major research program that included both exhaustive comparative studies and a very active experimentation on food webs. Tree-holes and container habitats such as the fauna of *Nepenthes* pitcher plants are very suitable subjects for this work. Kitching's book *Food Webs and Container Habitats* (2000) develops the subject in far more ways that I can easily summarize here. I will return to these comparative studies later.

Kitching and I used the organisms that inhabit natural water-filled tree-holes in subtropical rainforest ecosystems in Australia to test the ideas on food chain length. Energy enters tree-holes in the form of plant and animal detritus that falls into habitat units from the rainforest canopy. To circumvent any problem associated with the physical variability in natural tree-holes, we used plastic containers as analogues of natural habitat units. The first experiment conducted in subtropical rainforest in southeastern Queensland showed that the majority of species that inhabit natural tree-holes colonized water-filled plastic containers into which a quantity of leaf litter had been added as a source of energy (Pimm & Kitching, 1987). We varied productivity by using half to four times the natural rates of leaf litter input in their experiment.

This magnitude of difference in productivity did not affect food chain length significantly. However, the establishment of the containers did. It took much longer for the predators to colonize than the detritivores. We argued that, under natural circumstances where environmental vagaries might often eliminate species from natural tree-holes, the lower resilience of the longer food chains might be the principal factor in limiting their length.

In the extreme, energy must limit food chain length—for if there is no energy there can be no species. Nonetheless, this does not mean that the magnitude of energy flow is the best predictor of food chain lengths over the range of energy flows observed in nature. In a second experiment, Burt Jenkins, R. L. Kitching, and I used a far larger range of productivities than is likely to ever be encountered by natural tree-holes (Jenkins et al., 1992).

The 'high' energy treatment consisted of an initial loading of 6 grams of crushed leaf litter per container. This treatment also involved the input of subsequent installments of 0.6 grams of litter every six weeks. This is close to the average amount that we expected to enter a container over such a period if the container received only natural leaf litter falls. The 'medium' energy treatment consisted of an initial loading of 0.6 grams and an installment of 0.06 grams every six weeks which is an order of magnitude less than the amount of leaf litter added to containers in the 'high' energy treatment. The 'low' energy treatment had only an initial input of 0.06 grams of litter per container, an amount two orders of magnitude less than the 'high' energy treatment and ten times less than the 'medium' energy treatment. We added no further installments to the low energy containers.

We collected samples at weeks 6, 12, 24, 36, and 48 and observed food webs with the highest number of species, trophic links, and the longest chains in the twenty-fourth week. The lower the energy input, the fewer the trophic links, food chain lengths, and species were in the containers. There was also a decline in species numbers and food chain length in the thirty-sixth week as a result of dry conditions in the forest during which natural tree-holes in the vicinity of the experiment dried out. The effect on food chain length was most marked in the most productive system. We concluded that while relatively long food chains were possible only in the most productive systems, these systems were especially vulnerable to external perturbations.

A Null Hypothesis for Food Webs and Cohen's Cascade Model

Pages 124–30 deal with what I called the null hypothesis for food webs. It was an attempt to explore what food web features one might expect by chance, given the necessary constraints on an observed food web—its numbers of species of top-predators, of species that are both predators and prey, of species that are only prey, and the number of trophic interactions between them. I further constrained the webs so that there could be no trophic cycles (see above). Cohen's cascade model is very similar to this—a close cousin—and I shall return to his model soon.

Omnivory

Chapters 7, 8, and the first parts of Chapter 9 (pp. 131–76) deal with a variety of statistics that completed the set of patterns that one can derive from the food web itself. Of these, only omnivory has caused much controversy in the intervening years. I claimed that it was rarer than it should be, but there were a list of exceptions. These included webs dominated by parasitoids, detritivores that feed on the dead, and, after *Food Webs* was published, the species that feed their way up the food chain as they become larger. All of these are permissible exceptions in light of the unifying theory that nature abhors a food web that is likely to be unstable.

Over the two decades since these claims there have been many who have presented their webs, claiming them to show an abundance of omnivores. (See for example Williams & Martinez [2000] and the references therein.) Not once have I been convinced. *Food Webs* does not claim that there is "little omnivory" but simply that omnivores were rarer than expected by chance. "Read the instructions!" I often felt like writing. To my knowledge, no one has yet applied a food web null hypothesis (as I did) and shown that omnivores are as common (or more common) than one would expect. That doing this might be extremely difficult computationally for a web as complex as the one for Little Rock Lake (Martinez, 1991) is not an excuse for assuming the results are already in hand.

Nor have I ever understood why this pattern proved to be so contentious. Most ecologists routinely accept the idea that competition restricts the coexistence of two species, A and B. If A and B share a prey species, C, then they likely compete. So, if B makes A's life miserable by eating C, how much more so than if B also eats A! Simply, being both the prey and the competitor of another species is a tough option in a constantly changing world.

As a side note, feeding on different trophic levels in different food chains does not impose such dynamical constraints. The models do not exclude my eating potatoes and beef, since cows eat grass and not potatoes. *Food Webs* makes a clear distinction between within-chain omnivory (previous paragraph) and between-chain omnivory (this paragraph). Similarly, species that feed on carrion do not affect the abundance of their now-dead prey. Finally, species that eat their way up food chains may appear to feed on several trophic levels, but it is their individual life stages that feed on particular levels.

Compartments

The issue of compartments has a murky history. May (1973) suggested trophic interactions should be clumped, but his numerical argument was flawed (p. 144). Perhaps more importantly, he absorbed the conventional wisdom of ecologists of the day. Lawton and I only found evidence of compartments across broad divisions of habitats—as between the land and the sea, for example.

Even that barrier might not be what it seems. Gary Polis berated his fellow ecologists as they blithely assumed an ability to draw distinct boundaries around their systems. His studies of islands in the Sea of Cortez show that many of these islands received large amounts of energy and nutrients from the seas around them in the forms of bird droppings and washed-up carcasses. Islands, Michael Rose and he argued, need not be insular (Rose & Polis, 2000).

Predator–Prey Ratios

The topic of the ratios of predators to prey species in food webs was first raised by Cohen (1978). I criticized his specific ratio because the data available to Cohen were severely prejudiced against plants and invertebrates (p. 168). The top-predators of many food webs are described as species by the ecologists who reported them, but the lowest trophic level is often described as "plants," "detritus," or "phytoplankton." Jeffries & Lawton (1985) addressed the criticisms with a new data set of almost one hundred studies of freshwater habitats assembled with special care taken to address the taxonomic difficulties. They found a predator—prey ratio that varied from 1:2 in communities with few predators and prey species to 1:3.5 in communities with many predators and prey species. They found no differences in the ratio among communities in streams, rivers, or lakes, though across a wide range of communities the ratio is likely to vary. It is not the value of the ratio itself that is perhaps so interesting but its constancy within any broadly defined habitat.

Features of the predator overlap graphs

The final section of Chapter 9 (following p. 176) reviews the Cohen's book (1978). There is a distinction here that I did not emphasize as clearly as I would like to now. Food webs are a kind of graph, in the mathematical sense of that word. That is, a graph shows the relationships between objects. The objects in a food web are the species, and the relationship is a trophic one: A eats B. The food web is the basic graph, but one can derive other graphs from it.

One of these is the *predator overlap graph*—the objects are now the predators and the relationship involves the question of whether they share prey species, or henceforth, whether the predators overlap in their diet. To produce these graphs, we draw lines connecting each pair of predators that share one or more species of prey. These graphs show dietary overlap and dietary overlap indicates the potential for indirect (i.e., exploitive) competition.

In nature, predator overlap graphs show surprising regularities (Fig. F.2). Consider an approximate physical analogy. Make a physical model of the predator overlap graph with spheres to represent the predators and rods to connect those predators that share prey. The physical model of Fig. F.2(a) would be rigid, not flexible, because when there are connections around four or more predators these connections are triangulated. This would not be the case for the overlap graph in Fig. F.2(c). Also notice that species 8 in Fig. F.2(a) does not violate this condition because it is not part of a circuit around four or more species. The two unconnected parts of the graph (species 1–3 and 4–8) do not violate the condition for the same reason. If the connections from species 4–7 and 5–6 were missing, then the graph would not be triangulated. Technically, this property of being triangulated is called a *rigid circuit*. The predator overlap graphs of real food webs contain an overwhelming preponderance of rigid circuits compared to sets of computer- generated model food webs (Sugihara, 1984).

The pattern discussed by Cohen (and so in my Chapter 9, p. 176 and following) also comes from considering the overlap in the predators' use of their prey species. Cohen deemed a food web to be *interval* if the overlaps in the predators' use of prey species can be expressed as possibly overlapping segments of a line (Cohen, 1978).

For many arrangements of predator overlap graphs, the rigid circuit property ensures that the overlaps will have an interval representation. Yet the rigid circuit property does not *guarantee* that food webs are interval. It is possible to draw a non-interval, rigid circuit predator overlap graph, and Fig. F.2(b) is an example. The graph is a rigid circuit because there are no circuits around four or more points, yet it is still not possible to express the overlaps as overlapping segments of a line. For obvious reasons, this pattern is called *asteroidal*. If the overlap graph is not asteroidal and it is rigid circuit, it will be interval (Sugihara, 1984). Indeed, when real food webs are non-interval, it is because they are asteroidal.

Assembling the Prey Overlap Graphs

Another graph one can produce from the food web lets us combine information on both predators and their prey. One forms this graph by connecting prey species sharing a particular predator. Two prey species sharing a particular predator

(a)

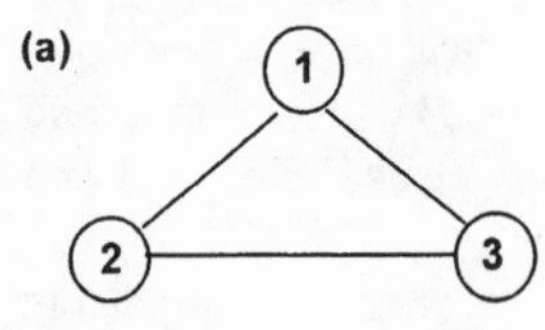

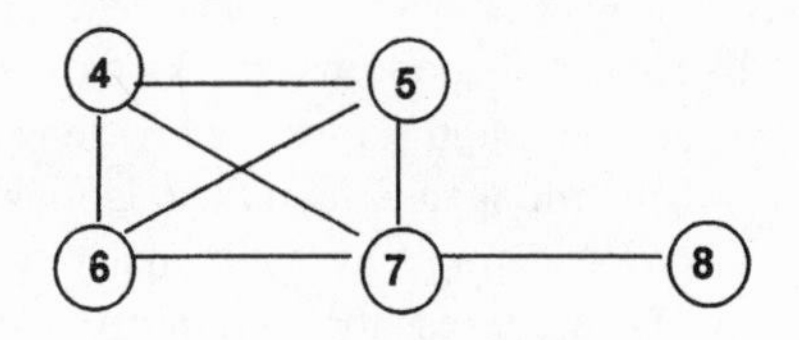

(b)

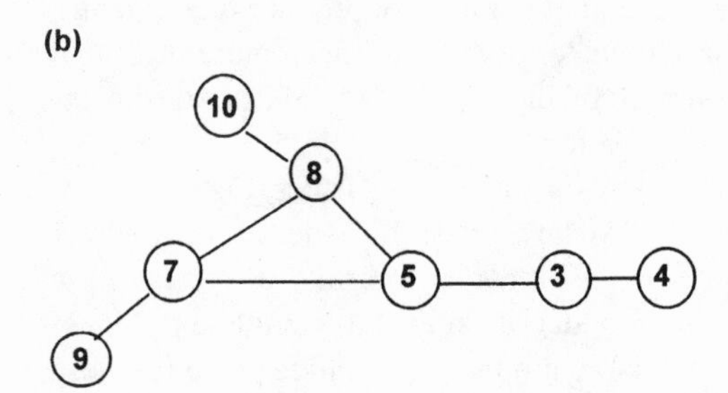

(c)

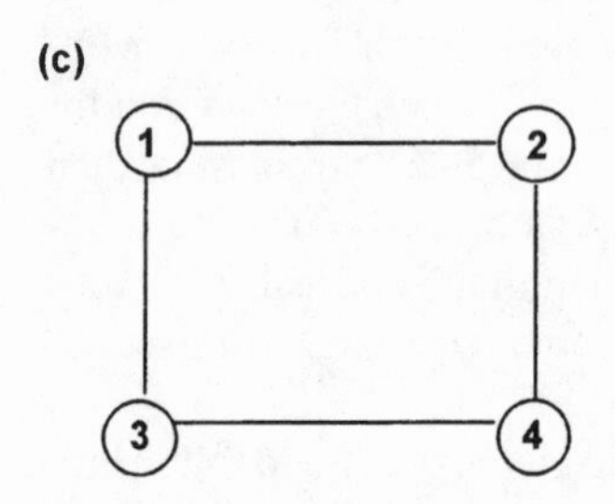

Figure F.2. Predator overlap graphs connect predators that share one or more prey species. Both (a) and (b) are rigid circuit, (c) is not. Only the predator overlap graph for (a) has an interval representation, as (b) is asteroidal (see text for definitions). Key: (a) is from Bird's (1930) study of a Canadian willow forest. 1, a fungus, 2, insects, 3, another group of insects, 4, three species of birds, 5, another three species of bird, 6, spiders, 7, a frog, 8, garter snake. (b) is from Kohn's (1959) study of predatory gastropods of the genus *Conus* on the sub-tidal reefs of Hawaii. All are species of *Conus*; 3 *ebraeus*, 4, *chaldeus,* 5, *miles,* 6, *rattus*, 7, *distans,* 8, *vexillum*, 9, *vitulinus,* 10, *imperialis.* (c) This pattern is so rare in nature that I have no example: this graph is hypothetical. (Pimm, 1991).

would form a line, three a triangular plane, four a tetrahedron, and so on. The graphs formed in this way are the *prey overlap graphs*.

Some prey species will feed more than one predator and this allows us to connect the individual graphs. Figure F.3(a) shows a combination of the graphs for three predators and how they exploit six species of prey. Figure F.3(b) is the food web from which I produce the prey overlap graph. Generally, I have used three prey species in the diet of each predator because planes are easier to draw than a multi-sided solid and so the examples are easier to visualize. Again, consider a

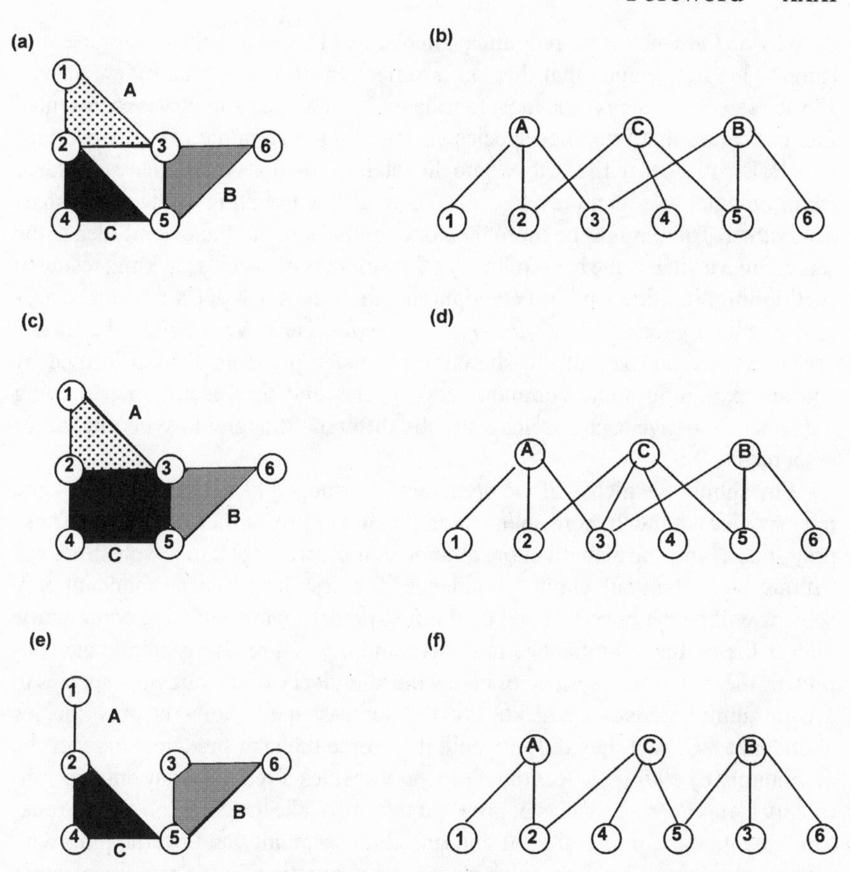

Figure F.3. Parts (a), (c), and (e) are the assembled prey overlap graphs, formed by connecting prey species that share a predator and parts (b), (d), and (f) are the food webs from which these graphs are derived. In (a) prey species **1**, **2**, and **3** are connected because predator ***A*** feeds on all three species. Species **1** and **4** are not connected because they have no predator in common. Prey connected in this way form "solids". (These are planes in all cases except the tetrahedron connecting **2**, **3**, **4**, and **5** in (c) and the line connecting **1** and **2** in (e).) This recipe of connecting species forms a hole between **2**, **3**, and **5** in (b). I discuss the significance of the different figures in the text. (Pimm, 1991.)

physical analogy of the overlap graph. The planes or the multi-sided objects formed by connecting the various prey species that share a particular predator are solid in the physical analogy. Continuing the physical analogy, observe that there is a hole between prey species 2, 3, and 5; no predator species feeds on this set of prey species. For another example, consider the web in Fig. F.3(c). Predator C now feeds on prey species 3 and so there is no longer a hole. These topological holes are rare in real food webs (Sugihara, 1984).

Why do these patterns predominate in Nature? Discussing them a decade later (Pimm, 1991), I argued that they are a reflection of an important view of how species select resources. The most familiar view of niche geometry is represented in a caricature that imagines species strung along a resource axis. (A overlaps with B, B with C, C with D, etc. where the relative positions might represent some environmental gradient, prey size, or a more abstract ranking of the prey's characteristics.) There might be more than one dimension, but the overall idea is the same. Such a view is the basis of many of the models of species packing, resource partitioning, limiting similarity, and many other ideas. I suggest a better caricature of niche geometry, the *flower petal model.* Here we imagine the niches drawn, as a Venn diagram, to show the extensive predator overlap formed by species exploiting some common prey species and the resource partitioning which tends to give each species a slightly different, though idiosyncratic, set of resources.

Why should the niches of the predators be arranged like flower petals? Some prey species are much more abundant in the diet of predators than others. (These prey species may be actually more abundant, more available, relatively more nutritious, etc. I shall talk about "abundance" for simplicity.) These abundant prey species will be the ones selected by the first predators to invade the community. Unless the predator's influence in the community is already so great that it depresses the common prey species below the abundance of the rare prey species in a community, predators will not create "donut-shaped" holes in prey species abundances. Only if they did so would they force the later predators to enter the community by taking a selection of rare prey species. Predators may enter a community by taking rare species of prey, but they also take the common ones. Predators thus give the impression of entering the community and overlapping with others predators where there are more species competing for the resources, rather than where there are few species sharing those resources. In short, it is the range of prey abundances that impose a ranking on the prey species so that the predators appear to avoid the ends of resource axes. I argue that this results in the topological regularities we have been discussing.

Small Worlds

That some species form the key centers for the action in the food web—those that are the cause of many species overlapping in the "flower petal" model just discussed—brings up my final topic about food web structure. First, a short preamble.

From the very start of my interest in food webs, it was obvious that many other things vaguely looked like food webs, and we could represent yet others as webs and the other graphs derived from them. As Steven Strogatz put it in his review of complex networks:

> The study of networks pervades all of science, from neurobiology to statistical physics. The most basic issues are structural: how does one characterize the

(ii) Cohen's work had shown that the average proportion of top-predators, intermediate species, and basal species remains roughly constant (but with high variance) in webs with widely differing numbers of species and from different habitats. The average proportion of trophic links that are between intermediate and intermediate species, intermediate species and top-predators, basal species and intermediate species, and basal species and top-predators remains constant (with large variance) in webs with widely differing numbers of species and from different habitats (Cohen et al., 1990).

Food Chain Length

Chapter 6 deals with food chain lengths. I still find it to be a useful introduction, but there is one important argument that I overlooked. Area must play an important role in limiting food chain lengths. In the limit, very small islands simply cannot have the production base to support top-predators. This is an important—and testable—extension of the energy flow argument developed on pages 104–10. Those pages cast doubt on the energy flow argument for short food chains because systems with low primary production do not obviously have shorter food chains than those with high production. In the former, the top-predators simply feed over larger areas. The area-limitation hypothesis simply asks what happens when area runs out! While per area production varies only over a couple of orders of magnitude, it's possible to compare the trophic levels of islands that span just a few square meters to those of more than 100,000 square kilometers. Not surprisingly, small islands have shorter food chain lengths (Schoener, 1989).

The argument that long food chains produce species with long recovery times (pp. 115–20) received support from an experimental study by Steven Carpenter and his colleagues. Carpenter et al. (1992) chose a small (area, 1.2ha), steep-sided (depth, 18.5m) experimental lake in Wisconsin and estimated both the flows and the stocks of phosphorous. Phosphorous is often a limiting nutrient in lakes. They aggregated the phosphorous in the web into six compartments: dissolved phosphorous, seston (mainly algae, but also the associated bacteria and protozoa), herbivorous zooplankton, *Chaoborus* (a predatory midge that feeds on the herbivores), planktivorous fish (that also feed on herbivores and also on the *Chaoborus*), and piscivorous fish. In 1984, planktivorous minnows were at the top of the food web. During 1985, over 50kg of minnows were removed and replaced by a similar mass of piscivorous largemouth bass. This re-configured the food web, adding an extra trophic level without significantly changing the total amount of phosphorous.

The team estimated the flows of phosphorous between compartments from the consumption of one by another. Other inputs to compartments include the emergence of eggs from benthic sediments. In both years, a major input was from the fish feeding in the lake's small but important littoral zone. There were flows of phosphorous back into the water column and losses, mainly to the benthic sediments. When all the numbers were in, neither year looks very much like the theo-

have some, and the occasional one has very many. So yes, you can connect the hermit in his cave to absolutely everyone provided the cave is in a congressional district where the currently serving member is running a tight race—and so wants to be friends with everyone, including the hermit.

How does the frequency of connections scale in real food webs? I don't know in general, though I do know the place to look—the species-rich webs described by Baird & Ulanowicz (1989), Goldwasser & Roughgarden (1993), Hall & Raffaelli (1991), Havens (1992), Martinez (1991), Polis (1991) Reagan & Waide (1996), and Warren (1989). Using the penultimate reference, Fig. F.4 shows the cumulative percentage of about 1,500 trophic interactions between 128 species of predators and their prey. The twenty-three most specialized species have only one prey species, seven more have only two prey species and these thirty species combined account for only about 2% of the trophic interactions. Thereafter the relationship accelerates. The twenty-two most generalized predators account for over half of all the trophic interactions. As in other networks, a few (nodes, people, internet web sites) are well connected, but most are not. How this scaling affects the propagation of disturbances through a food web has not yet been determined.

THE CAUSES OF FOOD WEB PATTERNS

Chapter 10 summarizes the list of food web patterns—a list to which Joel Cohen, John Lawton, Gary Polis, George Sugihara, and others greatly extended and for which Cohen, Lawton, and I were able to review in our 1991 paper (Pimm et al., 1991).

We asked: Are food web patterns artifacts? There were good reasons for concern about the quality of data in published webs. Communities often contain thousands of species. Because published webs include only tens of trophic species, they are either highly aggregated or represent only a tiny part of the entire system. Aggregation is rife in many published webs; moreover, aggregation varies in extent from web to web and at different positions in the same web. Even when webs are detailed enough for most of their elements to be single biological species, the linkages are less often based on experimental evidence than on casual observations. While accepting these problems, we also concluded that the evidence overwhelmingly rejected the patterns being artifactual.

Much has changed in the last decade as with numerous studies paying particular attention to the problems we raised. Some of these involve complex food webs, such as the linkages in food webs centered on gall wasps and, in particular, how alien species fit into new communities (Schönrogge & Crawley, 2001). Douglas Reagan and Robert Waide tackle perhaps one of the most complex food webs of all—that of a tropical rain forest (Reagan & Waide, 1996).

If food webs are patterned, then how many independent web patterns are there? Which patterns are the consequences of others? What causes the patterns? Of the various ideas put forward, two require particular comments—the cascade model and web dynamics.

Cohen's cascade model focused on the static patterns of trophic interaction and assigns linkages at random that are subject to two constraints (Cohen et al., 1990). First, the model assumed that we can arrange the species a priori into a cascade or hierarchy such that a given species can feed on only species below it, and itself can be fed on only by species above it in the hierarchy. This ordering automatically precludes trophic cycles and decomposer loops. It does not specify whether any particular species must be top, intermediate, or basal (except the lowest and highest species in the cascade). Second, the model requires two parameters obtained empirically: the number of species and the linkage density. By assumption, connectance declines hyperbolically.

By assigning linkages randomly within these constraints, the cascade model generates quantitative predictions that we can compare rigorously with observed patterns. It correctly predicts the average and variance of the fractions of all species that are basal, intermediate, and top-predators; the average fractions of linkages that are basal-intermediate, basal-top, intermediate-intermediate, and intermediate-top; the modal length of chains from basal to top species; and the decline in the frequencies of interval and rigid-circuit predator overlap graphs as webs get larger.

The cascade model was not used to explore some features of webs such as omnivory, compartments, and the ratios of how many prey species a species exploits to how many predatory species that species suffers. It also gets some of the fine details wrong though, again, problems in the quality of the data may be partly responsible for the discrepancy. For example, the predicted frequencies of very short and very long food chains within a given web are too high. The assumption of constant linkage density is challenged by data on species-rich webs (see above).

The original formulation of the cascade model offered no explanation for the postulated trophic cascade. Body size is the likely candidate because predators are typically larger than their prey and parasites are smaller (Cohen et al., 1993). The success of the cascade model is that it shows that many food web patterns are the consequence of a few simple underlying assumptions. It begs important questions.

What determines the linkage density? Quantitative theory is used to explain why the average species apparently utilizes, and is utilized by, a predictable and fairly small number of other species, and it is crucial to a deeper understanding and ultimate testing of the model. One explanation is the dynamics of the webs. Most webs are static descriptions, but the communities they describe are not static—as *Food Webs* asserts from start to finish.

With a decade of hindsight, I chose to reformulate the explanations of dynamics in terms of food web assembly. Some species are successful at invading a community while others are not, and the successes may or may not cause extinctions of former residents. This process of assembly and disintegration may explain many of the empirical web patterns (Pimm, 1991, Chapter 10). Some food web structures are hard to invade—they will persist. Others are easy to invade—they will not persist. In yet others, their instability means that some species will make quick exits.

In short, web dynamics is not an explanation incompatible with the cascade model. Rather, it suggests general mechanisms that limit linkage density and species richness and makes specific predictions about the details of web patterns.

THE RISE OF COMPARATIVE AND EXPERIMENTAL FOOD WEB ECOLOGY

Perhaps the most surprising fact found in *Food Webs* is how far one can take simple food web models and the empirical data on food webs (so often collected for every purpose but comparison) and blend them into a cohesive whole. The explanation is not that the models are sophisticated and the data excellent, but that the processes are basic and powerful and the patterns obvious and ubiquitous. Nonetheless, further progress would require far better data and careful, well-documented studies that compared food webs over time and space. Crucially, there would need to be thoughtful experiments to confirm the insights of these comparisons.

In the last twenty years, there have been many experiments on particular systems: the 1992 Carpenter et al. study is one. Bob Paine's long experience with inter-tidal communities suggested another approach—documenting the strengths of the interactions between species. Such strengths are so conspicuously absent from most representations of food webs! Paine (1992) found that most interactions were quite weak ones interspersed with a few strong ones.

It was obvious even when *Food Webs* first appeared that container habitats—tree-holes, the contents of pitcher plants, and so on—would allow both comparative and experimental studies of food web structure. Small, self-contained, highly replicated, and technically simple—the preferred tool for sampling them is a turkey baster—these *phytotelmata* were ideally suited to the task.

If I can be permitted one personal recollection, it would be of sitting in a undergraduate lab at Oxford in a white lab coat sorting horribly messy tree-hole samples from Wytham Wood for a practical led by Roger Kitching, then a graduate student. I recall Charles Elton coming in. I think he said something to the effect of "ecologists don't wear lab coats" though perhaps this is just part of the legend. Certainly, I have never worn one since, nor do I allow them in my lab. (The offending article still hangs in a wardrobe in my parents' house.) So, yes, I remembered Roger when I met him a decade later at a conference.

In a few paragraphs, I cannot do justice to his excellent new book *Food Webs and Container Habitats: The Natural History and Ecology of Phytotelmata* (Kitching, 2000). It is a superb natural history of these communities—Elton would be proud of his former graduate student—and a 100 pages are devoted to describing "the phytotelm bestiary." Almost another 100 pages describe the different kinds of container habitats and their environments. Thus armed, Kitching neatly overcomes most of the uncertainties of previous food web studies.

Better than any other study, Kitching demonstrates how gradually increasing spatial scale alters the explanations for food web structure. One of the best insights is offered by a figure that shows the local food web template—all the species and their potential trophic interactions at the location. A particular tree-hole's food web is "ghosted" onto this. A particular tree-hole will only have a subset of the species (and so their interactions), though this subset will change both seasonally and capriciously. The local food web itself is constrained by what species are present regionally. Particularly, for the food webs inside the pitcher plants of the genus *Nepenthes*, the Old World tropical distribution of species centered on Malaysia affects the number of coexisting species. That number also affects the number of species that might be found within an individual pitcher. In short, there are progressively larger scales that set the possible food webs that may be present in individual holes, in a food, or in a region, and there are factors—of the kind discussed here—that effect whether the full web will be realized.

CONSEQUENCES OF WEB PATTERNS

If food webs have structure, then so what? "Why is network anatomy so important to characterize?" Strogatz asks. "Because structure always affects function," he answers. That answer lies in *Food Webs,* and I lifted it from Charles Elton (1958).

Earlier in this foreword, I asked you to skip parts of *Food Webs.* In writing those parts, I struggled to define "stability" in some operational way. In addition to local dynamical stability, I suggested "species deletion stability," and I mentioned various experiments. None of this was at all satisfactory. Even less so were those couple of pages in Elton (1958) where he asserted that more complex systems would be more stable—all on the basis of a few completely unrelated anecdotes. The answer, of course, was that the recipe of stability outlined in *Food Webs* separated systems that would likely persist in Nature from those that would not. It was silent about how we might compare natural systems. The answer came from the recognition that "stability" also meant other things—resilience, variability, persistence, and resistance—and that these measures could applied to different variables—species abundances, total biomass, species composition. Elton's anecdotes were perceptive, but very incomplete. It wasn't until I understood the full range of meanings of stability that I could appreciate the connections he was trying to draw. It took me a decade before I could present the larger case that food web structure affects population and community dynamics in my next book *The Balance of Nature? Ecological Issues in the Conservation of Species and Communities* (Pimm, 1991). The arguments presented there are still hotly debated—the role of species numbers in affecting how resistant species are to change being the most recently controversial. (The papers of Tilman et al., 1997, Hector et al., 1999, and Huston et al., 2000 are sufficient to give the flavor of this debate.) And that's just on the number of species.

How connected food webs are, how long their food chains are—and so on down the list—are all factors that likely affect how natural communities change and respond to changes. In a world where we are changing so much so quickly, understanding natural complexity—what it means, how we are simplifying it, and what will be the consequences—is a manifestly important task.

References

Ariño, A. & Pimm, S. L. (1995), On the nature of population extremes. *Evolutionary Ecology,* **9**, 429–43.

Bird, R. P. (1930), Biotic communities of the aspen parkland of central Canada. *Ecology,* **11**, 356–442.

Baird, D. & Ulanowicz, R. E. (1989), The seasonal dynamics of the Chesapeake Bay ecosystem. *Ecological Monographs*, **59**, 329-64.

Bjørnstad, O. N. & Grenfell, B. T. (2001), Noisy clockwork: time series analysis of population fluctuations in animals. *Science*, **293**, 638–43

Carpenter, S. R., Kraft, C.E., Wright, R., He, Xi, Soranno, P.A., Hodgson, J.R: (1992), Resilience and resistance of a lake phosphorous cycle before and after food web manipulation. *The American Naturalist*, **140**, 781–98.

Cohen, J. E. (1978), *Food webs and niche space*, Princeton University Press, Princeton, New Jersey.

Cohen, J. E. (compiler) 1989a. Ecologists' Co-Operative Web Bank. ECOWeBTM Version 1.0. Machine-readable data base of food webs. New York: Rockefeller University.

Cohen, J. E., Briand, F., & Newman, C. M. (1990), *Community food webs: data and theory,* Springer-Verlag, New York.

Cohen, J. E., Pimm, S. L., Yodzis, P. & Saldaña, J. (1993), Body sizes of animal predators and animal prey in food webs. *Journal of Animal Ecology,* **62**, 67–78.

de Ruiter, P.C., Neutel, A. M., Moore, J. C. (1995), Emerging patterns of interaction strengths, and stability in real ecosystems. *Science* **269**, 1257-60.

Elton, C. S. (1958), *The Ecology of invasions by animals and plants*, Chapman & Hall, London.

Goldwasser, L. & Roughgarden, J. (1993), Construction of a large Caribbean food web. *Ecology,* **74**, 1216–33.

Hall, S. J.& Raffaelli, D. (1991), Food-web patterns: lessons from a species-rich web. *Journal of Animal Ecology,* **60**, 823–42.

Havens, K. (1992), Scale and structure in natural food webs. *Science,* **257**, 1107–9.

Hector, H. A. et al. (1999). Plant diversity and productivity in European grasslands. *Science,* **286**, 1123–27.

Hubbell, S. P. (2001), *The Unified Neutral Theory of Biodiversity and Biogeography,* Princeton University Press, Princeton, New Jersey.

Huston et al. (2000), No consistent effect of plant diversity on productivity. *Science,* **289**, 1255.

Inchausti, P. & Halley, J. (2001), Investigating long-term ecological variability using the global population dynamics database. *Nature,* **293**, 655–57

Jeffries, J. J. & Lawton, J. H. (1985), Predator—prey ratios in communities of freshwater invertebrates: the role of enemy free space. *Freshwater Biology,* **15**, 105–12.

Jenkins, B., Kitching, R. L. & Pimm, S. L. (1992), Productivity, disturbance, and food web structure at a local spatial scale in experimental container habitats. *Oikos,* **65**, 249–55.

Kitching, R. L. (2000), *Food Webs and Container Habitats,* Cambridge University Press, Cambridge.

Kohn, A. (1959), The ecology of *Conus* in Hawaii. *Ecological Monographs,* **29**, 47–90.

Lawton, J. H. (1988), More time means more variation. *Nature,* **334**, 563.

Manne, L. & Pimm, S. L. (1996), Ecology: engineered food webs. *Current Biology,* **6**, 29.

Martinez, N. D. (1991), Artifacts or attributes? Effects of resolution on the Little Rock Lake food web. *Ecological Monographs*, **61**, 367–92.

May, R. M. (1972), Will a large complex system be stable? *Nature,* **238**, 413–14.

May, R. M. (1973), *Stability and complexity in model ecosystems,* Princeton University Press, Princeton, New Jersey.

Paine, R. T. (1992), Food-web analysis through field measurement of per capita interaction strength. *Nature,* **355**, 73–75.

Pimm, S. L. (1991), *The Balance of Nature? Ecological issues in the conservation of species and communities,* The University of Chicago Press, Chicago.

Pimm, S. L. (2001), *The World According to Pimm: a Scientist Audits the Earth,* McGraw-Hill, New York.

Pimm, S. L. et al. (2001), Can we defy Nature's end? *Science,* **233**, 2207–20.

Pimm, S. L. & Kitching, R.L. (1987), The determinants of food chain lengths. *Oikos,* **50**, 302–7.

Pimm, S. L., Lawton, J. H. & Cohen, J. E. (1991), Food webs patterns and their consequences. *Nature,* **350**, 669–74.

Pimm, S. L. & Redfearn, A. (1988), The variability of animal populations. *Nature,* **334**, 613–14.

Pimm, S. L. & Rice, J. A. (1987), The dynamics of multispecies, multi-life-stage models of aquatic food webs. *Theoretical Population Biology,* **32**, 303–25.

Polis, G. A. (1991), Complex desert food webs: an empirical critique of food web theory. *The American Naturalist,* **138**, 123–55.

Reagan, D.P. & Waide, R. B. (eds.) (1996), *The Food Web of a Tropical Rain Forest,* The University of Chicago Press, Chicago.

Rose, M. D. & Polis, G. A. (2000), On the insularity of islands. *Ecography,* **23**, 693–701.

Schönrogge, K. & Crawley, M. J. (2001), Quantitative webs as a means of assessing the impact of alien insects. *Journal of Animal Ecology*, **69**, 841–68.

Strogatz, S. H. (2001), Exploring complex networks. *Nature,* **410**, 268–76.

Sugihara, G. (1984), Graph theory, homology and food webs. *Proceedings of Symposia in Applied Mathematics,* **30**, 83–101.

Tilman, D., Knops, J., Wedin, D., Reich, P., Ritchie, M. & Siemann, E. (1997), The influence of functional diversity and composition on ecosystem processes. *Science*, **277**, 1300–02.

Warren, P. H. (1989), Spatial and temporal variation in the structure of a freshwater food web. *Oikos,* **55**, 299–311.

Williams, R. J. & Martinez, N. D. (2000), Simple rules yield complex food webs. *Nature,* **440**, 181–83.

1 Food webs

1.1 WHAT AND WHY?

Food webs are diagrams depicting which species in a community interact. They depict binary relationships – whether species interact or not – and must miss much important biology. In the real world species interactions change at least seasonally and not all interactions are equally strong. Food webs are thus caricatures of nature. Like caricatures, though their representation of nature is distorted, there is enough truth to permit a study of some of the features they represent.

Food webs are complex and this has often been the reason for their appearance in the literature. They are often accompanied by expressions of the authors' despair at their complexity – a complexity which seems almost malevolent when it is necessary to explain it. But complexity does not mean randomness: I shall show that food webs are highly patterned. There is a repetition of features that prompts the discovery of the processes that generate them. By Chapter 10, a dozen patterns will have been discussed. The explanations for these patterns are fewer: constraints on population dynamics, energy flow, and the structural designs of animals explain them all. Simply, food webs are not the 'tangled knitting' or the 'spaghetti' they may resemble superficially. They possess structures shaped by a limited number of simple biological processes.

There are several reasons why food webs should be interesting. First, a catalogue of food web features has an intrinsic value independent of the factors that produce them. The catalogue provides a guide to what features should be incorporated into the ecosystem models which are playing an ever more important role in the understanding of applied and theoretical problems in ecology. Second, the explanation and prediction of food web patterns is central to the understanding of the processes at work in ecosystems. Finally, the structure of a food web critically affects the functions of the ecosystem to which it belongs. The resilience of an ecosystem to perturbations and how tightly an ecosystem retains its inorganic nutrients are examples of ecosystem functions dependent on food web design. As in other areas of biology, structure and function are intimately related.

1.2 WHERE?

Now, let me expand on these remarks by considering where in this book various arguments will appear. There are two general approaches to the question: how should food webs be studied? One approach involves assembling published food webs and closely scrutinizing them, usually with the aid of a computer. Thus patterns sometimes emerge and demand explanation. This approach is used in Chapter 9. I shall concentrate on another approach that uses ecological theory to predict which food web structures should be expected and then analyses the real webs to see if the theory is supported.

What theories should be used to predict food web structures? I shall use a recipe which has been remarkably successful in its predictions, but it would surprise me if there were not alternative approaches. The theory supposes that most populations are stable: populations persist despite constant perturbations to their densities. The final section of this chapter analyses data to show that this supposition is correct. Given stability, the question can be asked: which model food web structures are consistent with stability and which are not? The former, not the latter, are the structures predicted to be common in the real world.

Chapters 2 and 3 provide the mathematical basis for this recipe. First, it is necessary to decide on the models to be used to describe the populations' dynamics. There are two major formats. The first involves populations that change continuously through time (differential equation models), and the second involves populations where seasonality imposes short generation times and life histories make generations distinct (difference equation models). Within the former, two alternatives are recognized: where pedators affect the growth rate of their prey (Lotka–Volterra models) and where they do not (donor-controlled models). The donor-controlled models might be considered a special case of the Lotka–Volterra models, but their consequences to food web structures are so different that convenience demands a separate label for them.

Second, there are numerous definitions of stability. Different definitions imply different sizes or frequencies of perturbations. Moreover, exactly what is perturbed may differ: some authors model perturbations to the species' growth rates, while others model perturbations to the resources on which the species feed. This last consideration is critical because the two procedures result in diametrically opposite predictions about some food web features.

A problem should now be evident: with three model formats, two or more dimensions for perturbations, and the anticipation of a dozen or so food web features, the number of possible combinations is already too large to be investigated practically. Some pruning is essential. Chapter 3 provides some simplification of the definitions of stability: I note that not all the definitions make different predictions. In addition, the two major model formats suggest different taxa: difference equations are more appropriate for insects, differential equations for vertebrates. No decisions are drawn, however, as to whether species are modelled better by Lotka–Volterra or donor-controlled dynamics. Nor are there any conclusions about whether small or large perturbations are likely to be better at describing the processes that shape webs in nature. These decisions must await data that have a direct bearing on the topic of Chapters 4 and 5, namely food web complexity.

Chapter 4 presents a theoretical discussion of how web complexity is related to a number of definitions of stability. Early workers had supposed that the most complex systems were also the most stable, but more recent theoretical studies suggest the exact opposite. For example, using stability criteria involving only small perturbations, May (1972) found that increasing numbers of species, complexity, and the magnitude of interactions between species all decreased the likelihood of stability. Dynamical constraints place upper, not lower bounds on food web complexity. What are the consequences of a larger perturbation, say, entirely removing a species from a community? The consequences depend critically on whether Lotka–Volterra or donor-controlled dynamics are used. If the former are used, the results are comparable to May's: complexity decreases the chances of stability. If the latter are used, then the reverse is true.

There is an additional result: removals of only predators usually lead to further species losses with Lotka–Volterra models but have no effect (by definition) with donor-controlled models. It is this result that, in Chapter 5, allows some further simplification in the recipe for predicting food web structures.

Chapter 5 reviews some tests of the theories presented in Chapter 4. I argue that none of these tests is adequate to evaluate the validity of May's hypotheses. There is no consensus among studies that directly investigate the relationship between stability and complexity. Consequently, I suggest an indirect route. An assembly of studies in which species have been experimentally removed from communities shows that, nearly always, the consequences are further species losses. These results are incompatible with the assumption of donor control. And with the rejection of this model format go the otherwise most biologically plausible models where stability increases with complexity. Finally, the rejection of this format simplifies arguments in subsequent chapters; there is one less model format to be used in the prediction of food web structures. Moreover, because real systems cannot withstand perturbations as large as species removals, only relatively small perturbations are likely to shape webs selectively. In subsequent chapters, I use these results as a justification for concentrating on the effects of small perturbations to species densities modelled by Lotka–Volterra dynamics. The recipe for predicting food web structures now has sufficient simplicity to be useful.

Chapter 6 discusses the length of food chains which, typically, consist of three or four trophic levels. I review four major hypotheses for this limitation: (A) that, as the food chain is ascended, the rapidly attenuating energy flow precludes species feeding higher than at a certain level; (B) that it may be impossible to 'build' a predator to feed on existing top-predators; (C) that there are always advantages to feeding as low in the food chain as possible; and (D) that long food chains are dynamically fragile. Evidence suggests that hypothesis (A) cannot be correct except for systems with exceptionally low energy flow. All the remaining hypotheses are possibilities, though (B) is hard to test. Both (B) and (C) seem incompatible with a single study on aquatic communities that have three or four trophic levels in Australia but only two in England. The chemical and physical parameters of the habitats are almost identical and make it unlikely that either (B) or (C) could be possible for the English but not the Australian communities. If only by elimination, hypothesis (D) – dynamical constraints – seems the most likely contender.

Chapter 7 covers four patterns of web structures predicted by mathematical models. They involve patterns among omnivores, defined, for these purposes, as species that feed on more than one trophic level. The patterns are: (i) omnivores are relatively scarce, typically one per food chain; (ii) omnivores rarely exploit prey not at adjacent trophic levels; (iii) systems dominated by insects and their parasitoids permit much greater complexity in their patterns of omnivory than vertebrates (for which patterns (i) and (ii) hold); and (iv) donor-controlled systems also permit greater complexity in their patterns of omnivory. All four patterns are shown to be features of the real world.

Chapter 8 considers the question of compartments. Are species interactions grouped within webs so that interactions are strong (or numerous) within compartments but weak (or few) between compartments? Theoretical studies suggest the answer should be 'no'. Data show that there are compartments in the real world, but in places where simple biology would make us expect them. There is no evidence for compartments for purely dynamical reasons.

Chapter 9 discusses three food web features whose discovery comes from analysing real webs statistically: (i) The numbers of species of prey and of predators are in the ratio of 3:4. (ii) There is a negative correlation between the number of predatory species which a species suffers and the number of species of prey which the species exploits. (iii). The patterns of overlap in the species of prey exploited by a set of predators can usually be expressed in one dimension. The causes of the three patterns are uncertain, and the first and second are probably artifacts of the ways in which the data were recorded.

The last chapter reviews the dozen or so food web patterns and their causes and seeks a synthesis. The dynamical constraint of stability explains many but not all of the patterns. Some of the patterns explained by dynamics seem more likely to be consequences of simple biological constraints on animal feeding, while other patterns seem to have no satisfactory explanation other than dynamics. Simply, dynamical constraints appear to be necessary but not totally sufficient to explain the patterns in real food webs.

Finally, I address the relationships between web structure and ecosystem function. The discussion is brief, for a comprehensive coverage is not the objective. But what is presented shows the existence of important inter-relationships between structure and function.

Such is my prospectus. The remainder of this chapter will consider in more detail the methods to be used in subsequent chapters.

1.3 HOW?

1.3.1 Methods for studying food webs

The method on which I shall concentrate uses mathematical models to recreate the processes which shape webs. In the real world, systems persist despite the constant perturbations from the environment to the densities of their constituent species. Systems that cannot withstand such perturbations are assumed to be eliminated. What remains are food webs whose structures can be compared, statistically, with guesses at the set of 'all possible webs' from which they were derived (Fig. 1.1). Whether this idea of dynamical selection is correct or not, such statistical comparisons do indicate which features of food webs are non-random. Such straight statistical comparisons are the topic of Chapter 9. More interesting is the possibility of mimicking this process of dynamical selection by using models; this allows an anticipation of these comparisons. A large sample of random models is used initially and those that do not allow all their constituent species to persist are rejected. What remains is a final set of models with acceptable dynamics. Then the food webs of the final and initial sets can be compared in the hope of finding differences in their structures (Fig. 1.1). These differences provide predictions that can be tested

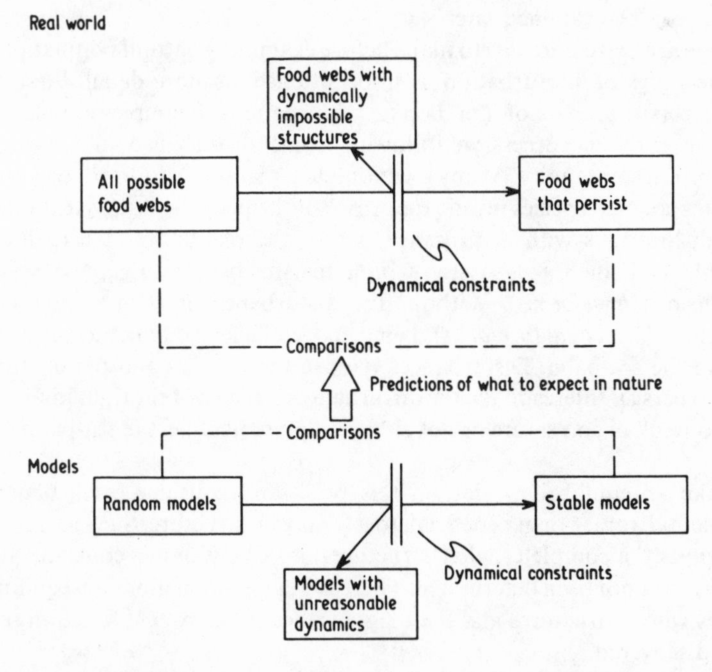

Figure 1.1 A scheme to predict food web structures from models. For further details see text.

against the statistical analyses of real food web features to see if the theory is acceptable or not.

The elimination of dynamically unreasonable structures is not the only process to shape food webs and it may not be the most important one. It is, however, one that has made some testable predictions about food web design that are supported by data from real webs. But the ease with which this tool can be applied may be deceptive: dynamical constraints make only ambiguous predictions about food web complexity (Chapters 4 and 5) and are only one of a number of reasons why the length of food chains may be limited (Chapter 6). Dynamical constraints do seem to be the best explanation for features involving omnivory (Chapter 7). Dynamics do not predict the patterns of compartments discussed in Chapter 8 nor the variety of features discussed in Chapter 9. In short, there is a warning: the emphasis on stability may be more a reflection of the content of current literature than on the relative importance of dynamics in restricting food web shape. It is premature to discuss this issue here. I shall return to it at the end of the book when more facts are at hand. Irrespective of the real world, I must develop both models and the machinery to analyse their stability in detail. Whether, in the long term, the predictions made from stability analyses are important or not does not deny that these predictions have been the starting point for most of the studies on food web shape. In short, stability analyses provide a convenient place to start; their failings can be examined later.

There are two comments to make before examining natural populations and the meanings of 'perturbation' and 'persistence' in more detail. First, many species persist *because* of, not *despite*, perturbations. Examples are plants that colonize early in succession following some disturbance to the original, equilibrium community. At any location, such plants are doomed to extinction as better adapted species invade the area. But the plants will survive if there are enough locations with disturbances which the plants are able to find. An example is Paine's (1979) study of a marine benthic alga, the sea-palm (*Postelsia palmaeformis*). Without the disturbance of strong wave action, mussels (*Mytilus californianus*), being more efficient competitors for space, remove the sea-palm. Disturbances remove the mussels and permit the sea-palm to persist. Interestingly, the disturbances must be of the right kind. If they are too weak or too severe or not sufficiently predictable, the sea-palm will be lost.

I take an equilibrium view in this book and consider perturbations as detrimental, rather than beneficial, to a food web structure. Such an approach is obviously incomplete; what structures are likely under constant perturbations have not been determined. There is a large number of non-equilibrium models whose structures and their significance to ecosystem function remain to be discovered.

The second comment involves the way webs are shaped by perturbations. If only those web structures that resist perturbations persist, a selection process is implied. Now, the process of natural selection acts upon the phenotypes of the individuals themselves. Nothing I have said implies any different kind of selection in shaping food webs. Specifically, the selection of food web patterns does not create some 'super-organism' with individuals consciously adopting strategies in a web's, but not their own, best interests. An example anticipates a result encountered in Chapter 7. Species feeding on more than one trophic level (omnivores) are relatively scarce. Such a species might be a carnivore that feeds on the plants on which its herbivorous prey also feed. The scarcity of omnivores does not mean there has been selection for carnivores that abstain from eating plants when it would be otherwise profitable for them to do so. Rather herbivores which not only suffer the attentions of a predator, but also lose substantial portions of their food to the same species, are likely to be driven to extinction.

1.3.2 Persistence and perturbations

I have presented a recipe which assumes that, although species densities are buffeted by perturbations, the species still persist. In this sense, the systems to which the species belong may be deemed 'stable' – though it will become apparent that definitions of stability must be made more exact than this if they are to be useful. Clearly, it must be made certain that the assumption of persistence is correct before proceeding.

To evaluate the assumption several questions must be answered. What exactly is meant by perturbation? Are perturbations large or small? Occasional or frequent? Transient or long-lasting? What is meant by persist? Do populations return to some initial equilibrium density following perturbation or do they persist at some other density? Do they continue to change in some repeatable, cyclical fashion? If they return to the initial density do they do so quickly, or slowly?

Model formats that are general enough to encompass all these possibilities are often mathematically intractable. More importantly, it is unlikely that all these possibilities are equally likely in shaping the structures of real food webs. If the model predictions are to be useful, the details of species persistence and environmental perturbations must also be correct.

For example, many real systems may not be able to withstand perturbations that are either too frequent or too severe. Models that withstand severe perturbations might be expected to have features of some real world systems, but so might many of the models that cannot withstand such severe shocks. Webs observed in nature may possess features that give both resistance and sensitivity to large perturbations in models. This will obscure the differences between the structures which may and may not be expected.

The other extreme may also be true: perturbations that are too weak may fail to eliminate those structures that cannot persist in the real world.

An analogy may help. Man designs buildings to withstand wind and rain. As a consequence, they have some obvious structural features (roofs, walls made of wood, brick or stone) that give them permanence in the face of the elements. A nomad's tent or a flimsy wooden shack will persist for a much shorter period. Of course, a tornado or an earthquake will destroy shack and stone edifice with equanimity, and a gentle breeze destroys neither.

In short, the recipe of 'persistence despite perturbations' is currently too vague to be useful and its arguments must be strengthened before they can be used to investigate food web structure.

Are populations stable?

Some insight into the actual patterns of persistence can be obtained by examining a range of population studies and asking whether species densities return to something like their original value following perturbations. Some clarification of this question comes from considering the patterns of dynamics exhibited by natural populations. Figure 1.2(a) shows the Song Thrush (*Turdus philomelus*) in English farmlands. The data are from the British Trust for Ornithology's Common Bird Census and are discussed in more detail in Chapter 3. Despite England's latitude, the winter climate is mild – but not always. During the winter of 1962–3 there was an unusually long period when the ground was frozen and covered with snow and ice. These conditions denied the thrushes access to their food supply and many birds starved. The breeding population in 1963 was considerably reduced. None the less, the population recovered within a few years, returning approximately to its original level. Despite the perturbation, the population persisted.

The pattern of density changes shown by the Song Thrush is common in the data on birds in English farmlands and woodlands. But is this pattern typical of other taxonomic groups and of populations elsewhere? Do all species have a clearly defined density about which they fluctuate and to which they return following some major perturbation? An examination of the range of population studies for unusual patterns shows that there is indeed a wide variation in the patterns shown by population densities. The Lynx (*Lynx canadensis*), of the Canadian Arctic, shows regular cyclical changes in abundance (Fig. 1.2(b), from Elton & Nicholson, 1942). *Bupulus piniarius*, an insect of the German forests, shows wide fluctuations over several orders of magnitude with no apparent pattern (Fig. 1.2(c), from Varley, 1949; note the logarithmic scale). Three more patterns come from the Common Bird Census data. Over long periods some populations like the Collared Dove (*Streptopelia decaocto*) or the Spotted Flycatcher (*Muscicapa striata*) show continuous increases or decreases in density (Figs. 1.2(d,e)). Finally, there are bizarre patterns like that of the Whitethroat (*Sylvia communis*), a small insectivorous bird common in hedgerows and which winters south of the Sahara. It appeared to fluctuate around one equilibrium before 1968 and another, much lower one, after that date (Fig. 1.2(f)). I have selected these

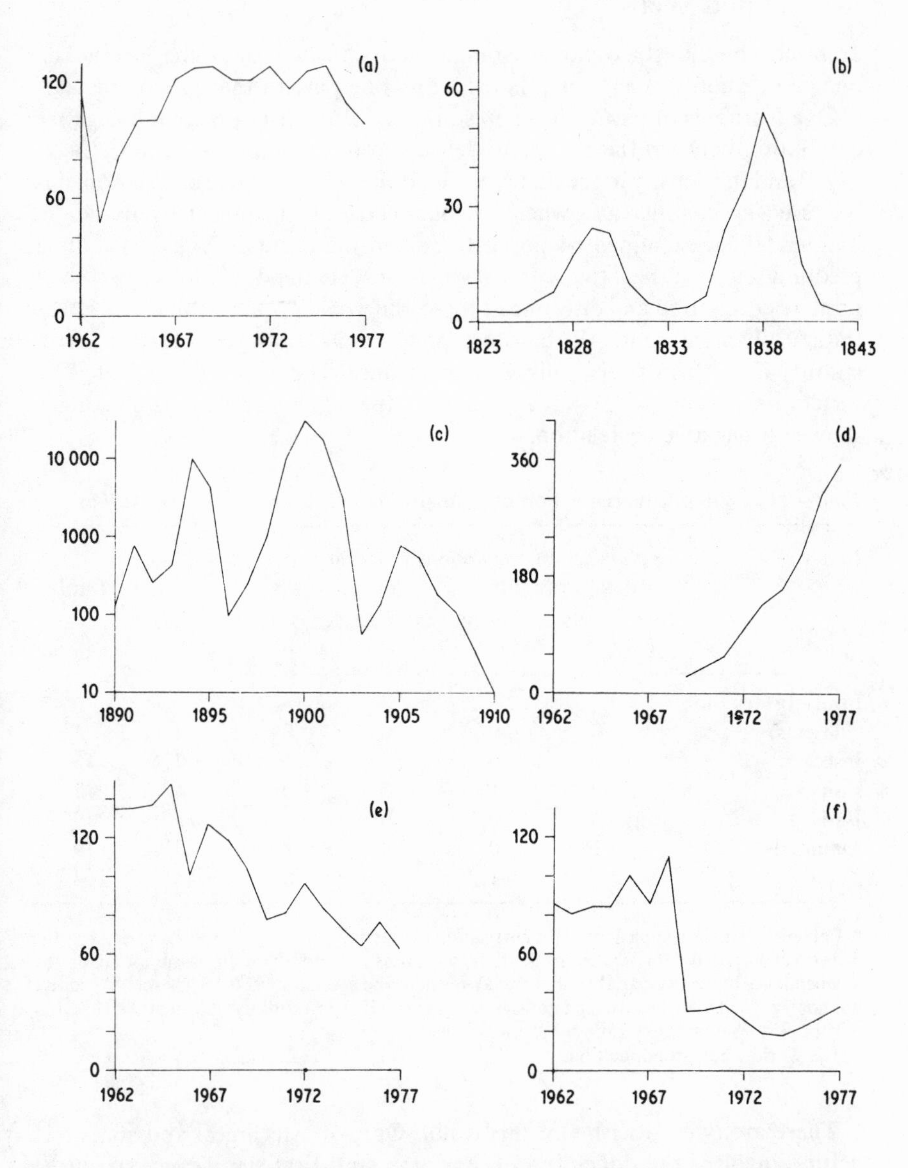

Figure 1.2 Some examples of population changes. (a) Song Thrush in English farmlands, (b) Lynx in the Canadian Arctic, (c) *Bupulus*, a moth in German forests, (d) Collared Dove in England, (e) Spotted Flycatcher in English farmlands, and (f) Whitethroat in English farmlands. Scales: the abscissa in all cases is in years. The units of density are relative ones in (a), (d), (e) and (f), and are set to 100 in either 1966 (a), (e) and (f) or in 1972, (d). In (b) the units are thousands trapped per year and in (c) numbers per hectare of forest floor. Sources are given in the text.

examples consciously to show the range of possible patterns; some, like the last one, are certainly exceptional. Is there any one pattern that predominates?

One feature is of particular interest. It is whether or not there is a negative correlation between the change in density from one year to the next ($X_{t+1} - X_t$) and the density in the first year, X_t. If there is a negative correlation, it is because a species increases when rare but decreases when unusually abundant. Tanner (1966) examined 64 populations and his results (Table 1.1) show a predominance of negative correlations. Only two populations showed a positive correlation and only one of these, the world's human population, was statistically significant. All the other populations displayed negative correlations, 42 of them statistically significant ones. The data are dominated by vertebrates (about 70%), but even among the insect populations all studies showed a negative correlation.

Table 1.1 Data on the regulation of animal populations. From Tanner (1966).

Taxon	*Numbers of populations in various categories*					
	*(a)** +*ve sig.*	*(b)* +*ve ns*	*(c)* −*ve ns*	*(d)* −*ve sig. at* 10%	*(e)* −*ve sig. at* 5%	*Total*
Invertebrates (not insects)	0	0	0	0	4	4
Insects	0	0	7	1	7	15
Fish	0	1	2	0	4	7
Birds	0	0	3	3	13	19
Mammals	1†	0	4	1	13	19
Total	1	1	16	5	41	64

* Categories are the significance of the correlation between $(X_{t+1} - X_t)/X_t$ and X_t, where X_i is the density of a population in the ith year. Significances are based on the assumption of the independence of the two variables and are (a) positive and significant at the 5% level, (b) positive but not significant, (c) negative and not significant at the 10% level, (d) significant at the 10%, but not the 5% level and (e) significant at the 5% level.
† The world's human population.

There are two criticisms of this result. One, of a technical and statistical nature, involves the difficulties of assigning statistical significance to correlations based on serially correlated data (Ito, 1972). I doubt whether this criticism is severe enough to reverse the overall conclusion. The second is that ecologists might prefer to study populations that are fairly stable over a long period of time. After all, a population that is here today but gone next year would be a singularly frustrating one on which to work. In answer to this, I analysed Common Bird Census data on 45 species of birds in English farmlands and woodlands (Pimm, 1982). Species were probably selected in this project by their abundance rather than by their perceived stability. The statistical analyses were similar to those of Tanner and are discussed later in a different context (Chapter 3). Most of the correlations between change and density (Table 1.2) were negative, although not all were statistically significant.

What is the biological significance of this predominance of negative correlations? Certainly, most natural populations tend to stay within bounds because of the density-dependent processes that reduce their numbers when high but permit them to increase when low. The majority of the populations, however, show *strong* negative correlations between change and density. Such populations do not include those with violent oscillations typical of species like *Bupulus piniarius*. Strong negative correlations tend to be found in species, like the Song Thrush, whose densities quickly return to a recognizable equilibrium. However, the cyclical pattern of densities shown by the Lynx also results in a strong negative correlation, though these cycles are unusual.

Table 1.2 Data on the regulation of the population densities of common birds in British woodlands and farmlands.

Habitat	*Numbers of populations in various categories*					
	*(a)**	*(b)*	*(c)*	*(d)*	*(e)*	*Total*
	+ve sig.	*+ve ns*	*−ve ns*	*−ve sig. at 10%*	*−ve sig. at 5%*	
Farmland species	0	0	13	7	22	42
Woodland species	0	2	16	6	8	32

* Categories are the significance of the term, β_2, in the regression equation $X_{t+1} = \beta_1 X_t + \beta_2 X_t^2$, where the β_i are estimated by a least squares procedure from the X_i, the density of a species in the ith year. The significance levels assume the variation about this model to be normally and independently distributed. The categories have the same designations as in Table 1.1. There are no column totals because many species are present in both farmland and woodland habitats.

To summarize, nearly all natural populations are characterized by patterns of change that keep their numbers within bounds. Moreover, the majority have strong negative correlations between change and density. This suggests that following a perturbation, densities tend to return to a recognizable equilibrium level or, more rarely, a cyclical pattern. Only a minority of populations fluctuate so wildly that an equilibrium level is not obvious. Most populations demonstrate the existence of processes that enhance their persistence. In some sense of the word, populations are stable. But the definition of stability I shall use is more restrictive: it is that population densities return to an equilibrium following a perturbation. It appears that the majority of populations also satisfy this more restrictive definition. Now I shall consider the ways of simulating population dynamics in mathematical models and of deciding whether the models satisfy this restrictive definition of stability.

2 Models and their local stability

2.1 INTRODUCTION

This chapter presents several models of population change and examines some simple but approximate methods for determining their stability. In the next chapter, I shall consider alternative and more complex stability criteria.

2.1.1 A note on mathematics

The mathematics treated in this and the next chapter usually occurs towards the end of introductory texts on differential equations. It also requires a knowledge of linear algebra. These are two factors which considerably reduce its accessibility to many biologists. Space prohibits a detailed and rigorous discussion with enough background to be suitable for those with limited experience in calculus. Concise and rigorous treatments of the mathematics are available in many texts on physics, mathematics, and engineering; it would be pointless to duplicate them. I shall instead present discussions of two-species models and their stability in sufficient detail to delineate the biological assumptions that must be made if the mathematics is to be tractable. In subsequent chapters emphasis will be placed on evaluating these assumptions, so it seems consistent to clarify why the assumptions are necessary in the first place. The mathematics I present requires only a limited knowledge of calculus. I then ask the reader to accept that the results for two-species models generalize to n-species models without any additional biological assumptions. For the biologist with a total aversion to mathematics, I have tried to make the chapter summaries sufficiently explanatory to make subsequent chapters intelligible. For those with mathematical backgrounds, this and the next chapter represent what May (1973a) has described as 'turning the handle of some well defined mathematical machinery'.

2.2 MODELS

I now present three model formats of population growth and interaction, and then discuss an approximate method for determining their stability. The first format, due largely to Lotka (1925) and Volterra (1926), has a predator's growth rate regulated by the abundance of its prey. The prey's growth rate is determined by its food supply as well as by the density of its predators. The second format has growth rates regulated solely by the abundance of the species' prey. The predators only 'skim off' the prey that are doomed to die from other causes and hence do not affect the growth rate of their prey. These two formats lead to different predictions about food web structure. There are examples of both in the real world, but data will show that the assumptions of the first format seem more general. In both formats birth and death processes are continuous. This ideal may be most closely approximated by species living in such relatively aseasonal environments as humid tropical forests. In the third format, models are produced for species that live in highly seasonal environments where species are dormant for part of the year.

There is a wealth of models to describe the detailed interactions between populations. I shall consider some very simple models in the hope that they capture some of the essence of the real world; the properties of more complex models are often impossible to analyse with insight. Still, only a tiny fraction of the possible models are considered. That this list is not exhaustive is not critical in itself. Many model features do not alter dynamics enough to reverse the qualitative predictions developed. For example, the emphasis throughout will be on Lotka–Volterra models. They are a convenient starting place and I shall frame many of my arguments in their context. Nearly all the results in Chapters 4–10 are based on the behaviour of species densities close to equilibrium. Whereas a system's behaviour after large perturbations may depend critically on the exact models used, a wide variety of models give approximations for small perturbations that are identical to the Lotka–Volterra models. Where model features do critically affect predictions, I have attempted to evaluate the models' predictions to decide whether the models are appropriate or not. The three formats discussed are among those that produce the most dissimilar predictions about food web shapes while retaining biological reality. They are also among the most widely studied formats.

2.2.1 Differential equations I: Lotka–Volterra

The aim of food web models seems disarmingly simple: there must be an equation for the growth rate of every species in the system. Each equation must include terms that reflect how this species would grow in the absence of all other species and how the other species affect its growth. The complexities of the models arise when *how* species interact is considered.

Suppose there are n species in the food web. A typical species is X_i, and this term will be used to refer to both the species and its density. It is fundamental to these models that the rate of change of X_i, called $\dot{X}_i$, is known rather than X_i directly. The term $\dot{X}_i$ is often written $\mathrm{d}X_i/\mathrm{d}t$, and the meanings are identical. In the absence of immigration and emigration, $\dot{X}_i$ will be the difference between the birth and death rates of the population.

(*a*) *Birth rates*
The quantity $\dot{X}_i/X_i$ is the *per capita* rate of change of the population. Consider the component of this that is due to births. There are two cases: a predator feeding upon a prey population and a prey population which is limited by some resource such as space or nutrients. Consider the predator (call it X_2) first: a simple assumption is that the birth rate per individual is directly proportional to the abundance of its prey (X_1). The more prey there are, the more an individual predator will consume and the more offspring it will produce as a consequence. Thus

$$\dot{X}_2/X_2 = a_{21}X_1, \tag{2.1}$$

where a_{21} is the number of predators produced per predator, for each prey present (of which only some will be consumed). Note that the subscripts on this and similar terms are of the form a_{ij} and indicate the effect species j has on the growth rate of species i. The birth rate of the prey is of a different form because it feeds on a resource whose availability is assumed constant. Because of this, assume that the birth rate of the prey is constant:

$$\dot{X}_1/X_1 = b_1. \tag{2.2}$$

(*b*) *Death rates*
Again there are two cases. The predator without a food supply is assumed to die off at a constant rate per individual:

$$\dot{X}_2/X_2 = b_2. \tag{2.3}$$

The death rate of the prey is more complex. As X_1 increases, the amount of resource available to each individual will decline and the death rate will increase:

$$\dot{X}_1/X_1 = a_{11}X_1. \tag{2.4}$$

This is not the only mortality the prey suffer because the predators also kill some of the prey. Assume that this mortality is proportional to the number of predators. The total death rate becomes

$$\dot{X}_1/X_1 = a_{11}X_1 + a_{12}X_2, \tag{2.5}$$

where a_{12} is the number of prey, per prey, killed by each predator. Subtracting the death rates from the birth rates yields

$$\begin{aligned}\dot{X}_1/X_1 &= b_1 - a_{11}X_1 - a_{12}X_2\\ \dot{X}_2/X_2 &= -b_2 + a_{21}X_1.\end{aligned} \tag{2.6}$$

Finally, to obtain the rate of change for the total population multiply each side of the equation by its corresponding population density (X_i):

$$\dot{X}_1 = X_1(b_1 - a_{11}X_1 - a_{12}X_2)$$
$$\dot{X}_2 = X_2(-b_2 + a_{21}X_1). \quad (2.7)$$

In a similar manner, a complex multispecies system can be modelled with the generalizations of Equations (2.7). For a species in a system of n species a typical equation can be written as

$$\dot{X}_i = X_i(b_i + \sum_{j=1}^{n} a_{ij}X_j), \quad (2.8)$$

where the sign and the magnitude of a_{ij} will depend on whether X_i is the predator and X_j the prey, or *vice versa*. If the two species do not interact directly then $a_{ij} = a_{ji} = 0$; such would be the case between species X_1 and X_3 in a simple food chain where species X_3 feeds on X_2 which feeds on X_1. *Thus, the sign structure of a_{ij} represents the connections within the food web.* Various combinations of signs for a_{ij} and a_{ji} are possible: for example, predator–prey $(+,-)$, competitive $(-,-)$, and mutualistic $(+,+)$ interactions can be modelled.

These models have features on which comment is unavoidable. The models may seem inconsistent because b_i are birth rates for prey but death rates for predators. The $a_{ij}X_j$ terms may also represent both birth and death rates. Biologically, the assumptions of the Lotka–Volterra models are rarely defended. Mathematically, the models are convenient because they contain terms that, *per capita*, are either constant (X^0, i.e. b_i) or contain first-order interactions (X^1, i.e. $a_{ij}X_j$). These are the first two terms of a polynomial function. Now, any complicated function $f(Y)$ can be approximated about a given point by a polynomial – a function of Y^0, Y^1, Y^2, Y^3, etc. with increasing accuracy as more terms are added. (This idea is discussed in more detail below and in Appendix 2A.) This result allows a defence of the Lotka–Volterra models even if the biological assumptions sketched above are incorrect. Suppose only some very complicated functions accurately describe the populations' growth rates. The Lotka–Volterra models may still be correct, but only in that they represent the first two terms of polynomial approximations of these functions. In such circumstances there may be no simple biological meaning to the parameters. Both birth and death rates may be complicated functions with constant and first-order terms (X^0, X^1) with, for example, the b_i representing the difference between the constant terms approximating the birth and death rates. This solves one problem, namely the apparent inconsistency of the meanings of the terms, but creates another: what values should the parameters hold? There is no easy solution to this new problem. Later, I shall adopt two strategies. The first is to argue that simple biology places certain sign constraints on the parameters and the second is to choose the magnitude of the parameters randomly over intervals which also reflect reasonable biological constraints. This latter strategy allows for both the natural variability of the parameters and the uncertainty about their values.

2.2.2 Differential equations II: donor control

An entirely different set of models results from the assumptions about the death rates of a prey species. A detailed and fascinating study of a predator and its prey is Kruuk's (1972) work on the Spotted Hyena (*Crocuta crocuta*) in the equatorial grasslands of East Africa. Though the hyenas hunt in packs and are formidable predators, they often select animals that seem most likely to die for other reasons, including starvation. Kruuk writes (p. 101):

> '. . . from all the three main prey categories hyenas select certain categories. Very young animals are obviously extremely vulnerable, as are very old Wildebeest and probably gazelle. The role of disease is virtually unknown; from direct observations hyenas appear to select sick animals if they are present . . . . The hyenas' selection and hunting methods cause them to select the least physiologically fit from the populations at least in Wildebeest and gazelle . . . . This by no means signifies that hyenas eat only very old, young or diseased animals, but if these are available they will be selected.'

Clearly, the hyenas prefer to feed on some of the animals that are on the verge of dying from lack of food. The old and the very young are often those that are most likely to be included in the resource-dependent death rate ($a_{11}X_1$) in the model developed above. Moreover, animals are much more likely to succumb to disease if they are starving. If these animals constitute the bulk of the animals that die from predation, then it is inconsistent to include two terms for the mortality of the prey in Equation (2.4). Consider the extreme case when the predators take only those animals that are bound to die from starvation. The predators take a resource (dying animals) over whose abundance they have, or at least exercise, no control. The predators should be able to increase when rare, for there will be a relative abundance of dying animals. The death rate of tne predators should depend on the ratio of their own numbers to the numbers of the dying animals, which in turn is dependent on the density of the prey. It would be more appropriate to model such a system with equations of the form

$$\dot{X}_1 = X_1(b_1 - a_{11}X_1)$$
$$\dot{X}_2 = X_2[b_2 - a_{21}(X_2/a_{11}X_1)]. \qquad (2.9)$$

There is no apparent predation term in the equation for X_1 because the predators are eating some, or all, of the prey dying from lack of resources ($a_{11}X_1$). The predators increase when their numbers are below a limit set by the ratio of their numbers to the availability of food (dying animals) multiplied by the rate at which they can convert these into new predators (a_{21}).

In contrast to the Lotka–Volterra models, the constant terms, b_i, represent birth rates in both predators and prey and, similarly, the a_{ij} terms are consistently death rates. Another contrast is in the effect that the predators have on the equilibrium density of their prey. In both sets of models this density (called X_1^*) is obtained by setting $\dot{X}_1 = 0$ (no change in population) and ignoring the so-called trivial equilibrium when $X_1 = 0$. Comparing Equations (2.7) with (2.9):

from (2.7) $$b_1 - a_{11}X_1^* - a_{12}X_2 = 0,$$
or $$X_1^* = (b_1 - a_{12}X_2)/a_{11}; \tag{2.10}$$
from (2.9) $$b_1 - a_{11}X_1^* = 0,$$
or $$X_1^* = b_1/a_{11}. \tag{2.11}$$

In the Lotka–Volterra models the equilibrium density of the prey depends, in part, on that of the predator. Under donor-controlled models this is not so: the donor (prey) controls the density of the recipient population (predator) but not the reverse. The distinction is important. Many predictions to be made about expected food web features differ between Lotka–Volterra and donor-controlled models. Such sensitivity of predictions to the models' assumptions is a serious problem. The solution comes from testable predictions of Equations (2.10) and (2.11). Removal of a predator will lead to a change in the equilibrium density of the prey if Lotka–Volterra, but not if donor-controlled, dynamics are the better description of nature. To anticipate a result of later chapters, removal of predators usually leads to significant changes in the density of their prey. Consequently, I shall concentrate on Lotka–Volterra models in the examples developed in this chapter and the next. Donor-controlled models will be discussed, however, in subsequent chapters, where their predictions are divergent from those of Lotka–Volterra models.

2.2.3 Difference equations

Predator–prey equations are usually presented in one of two forms: differential and difference equations. The differential equations are those examined so far. They are appropriate to life cycles where generations overlap and birth and death processes are continuous. This ideal is rarely encountered, although some populations in such aseasonal environments as humid tropical forests must come close. In contrast, difference equations are appropriate for describing populations characterized by density changes occurring at discrete time units, often the time it takes to complete one generation. In these equations, population densities depend on the population level at some previous time period separated by a finite time interval. These models include the time delays (of one time unit) that figure prominently in the real world but which are lacking from the differential equations discussed above. Difference equations are most appropriate to populations with distinct generations, for example, insects in temperate climates where diapause during the winter months is common. Most populations fall between the two extremes modelled by the differential and difference equation forms, so it is important to see whether both forms lead to similar predictions about food web structures. If this is so, more confidence can be expressed that the results are not due to an arbitrary choice of one of these model formats.

Our knowledge of the dynamics of difference equations is well summarized by Hassell (1979). His models deal with the interactions between parasitoids and their hosts. Parasitoids are insect predators with some special features. They invariably kill their hosts – unlike true parasites, but like predators. These insects belong mainly to the orders of Diptera and Hymenoptera. Thus they might seem too restricted in occurrence to warrant the attention given them. This is not so: not only are they economically important in the biological control of insect pests, but they comprise perhaps 14% of all known insects (Hassell, 1979).

The female parasitoid lays an egg on, in, or near its host sometimes after immobilizing it. The egg hatches and the larval parasitoid eventually consumes the host. These features make this interaction a delight to model. Unlike predators, where males and females, young and old all locate and consume prey, only the adult female parasitoid searches for prey. Only one equation is needed to describe this search. The number of hosts parasitized determines, quite simply, the number of subsequent parasitoid progeny; often only one parasitoid will be produced per host parasitized.

Considerable behavioural sophistication has been built into these models, but the simplest assumption is that the parasitoids search randomly. When parasitoids (X_2) search randomly over a specified area, the number of hosts (X_1) they encounter at time $t (X_{\mathrm{le},t})$ relative to the total available ($X_{1,t}$) increases as their numbers ($X_{2,t}$) increase. But as more hosts are found, it becomes much less likely that unparasitized ones will be encountered. The exact relationship is derived by Hassell (1979), and it is a familiar one to collectors of stamps, butterflies or bird species seen: the more one has collected, the harder it is to get new ones. Mathematically, the result is expressed by

$$X_{\mathrm{le},t}/X_{1,t} = 1 - \exp(-a_{12}X_{2,t}). \tag{2.12}$$

By exp (z) I mean a quantity e – the base of natural logarithms – raised to the power z. (In general, explanations of symbols are to be found in the 'Convention and Definition' section on page xi). The quantity a_{12} is a constant known as the 'area of search'; the larger the constant, the greater the proportion of hosts encountered, for a given density of parasitoids. If only one parasitoid hatches per host then the density of the parasitoids in the next generation, $X_{2,t+1}$, is the same as the density of hosts encountered, $X_{\mathrm{le},t}$:

$$X_{2,t+1} = X_{1,t}[1 - \exp(-a_{12}X_{2,t})]. \tag{2.13}$$

The density of hosts surviving to reproduce is, at best, $X_1 - X_{\mathrm{le}}$. Suppose that each surviving host produces b_1 new hosts. Then the number of hosts in the next generation, $X_{1,t+1}$, is

$$X_{1,t+1} = b_1(X_{1,t} - X_{\mathrm{le},t}) = b_1 X_{1,t}\exp(-a_{12}X_{2,t}). \tag{2.14}$$

Equations (2.13) and (2.14) are known as the Nicholson–Bailey equations after the authors who first investigated their behaviour (Nicholson & Bailey, 1935). The number of hosts that survive parasitism is likely to be more than survive to reproduce, as some will die from a shortage of food. Beddington, Free & Lawton (1975) suggested a difference equation to model this resource limitation:

$$X_{1,t+1} = X_{1,t} b_1 \exp(1 - a_{11} X_{1,t}). \tag{2.15}$$

This equation is an analogue of the birth and death equations discussed above in differential equation form. When $X_{1,t}$ is small, the *per capita* rate of increase depends on b_1. There is a density-dependent death rate (as in Equation (2.4)) increasing in magnitude so that at $X_{1,t} = (1 + \ln(b_1))/a_{11}$, $X_{1,t+1}$ will equal $X_{1,t}$. Combining the mortalities from parasitism and resource limitation, Beddington *et al.* obtained a final equation for the dynamics of the host:

$$X_{1,t+1} = X_{1,t} b_1 \exp(1 - a_{11} X_{1,t} - a_{12} X_{2,t}). \tag{2.16}$$

2.3 STABILITY

2.3.1 Introduction

Given these models, a knowledge of how they describe population changes through time is of critical importance. Specifically, do the population densities tend towards some constant level despite perturbations to the system? To answer this question requires a knowledge of stability. First, I shall consider a single-species model to illustrate some obvious ways of determining stability. These, perversely, will not work with complex models, but it is necessary to understand why they do not work in order to see why other methods must be used. These other methods will be discussed, in detail, presently.

Consider a simple single-species population model of a species limited by a resource and having no predator:

$$\dot{X}_1 = X_1(b_1 - a_{11} X_1). \tag{2.17}$$

There are two cases where the population is at equilibrium (i.e. $\dot{X}_1 = 0$): the first is the trivial equilibrium where $X_1 = 0$; and the second, the non-trivial equilibrium, is at $X_1 = b_1/a_{11}$. For all models, I call X_i^* the value of X_i where $\dot{X}_i = 0$ and $X_i \neq 0$. The quantity b_1/a_{11} is often called the carrying capacity, K, of a species. The growth rate, b_1, is often labelled r, leading to a more familiar representation of Equation (2.17):

$$\dot{X} = rX(1 - X/K). \tag{2.18}$$

I retain the notation of b_i and a_{ij} because it is more convenient for models with many species. How can it be decided whether Equation (2.18) is stable? These are three ways:

(i) By far the simplest method is to note that for any X_1 below X_1^* (other than $X_1 = 0$) the population density always increases and for any X_1 above X_1^* the population density always decreases. The model is stable because the population tends to equilibrium from any density.

(ii) Equation (2.17) can be integrated to obtain an explicit solution of X_1 as a function of time. Such an integration is of limited use in a discussion of food webs, so the result is given without details (it may be found in many introductory texts on population dynamics, e.g. Roughgarden, 1979, p. 304). The result is

$$X_t = [(b_1/a_{11})X_0 \exp(b_1 t)]/[X_0 \exp(b_1 t) + (b_1/a_{11}) - X_0]. \quad (2.19)$$

Here, X_0 is the initial density of X and X_t the density of X after a period of time t. As time increases, the dominant term in the denominator becomes $X_0 \exp(b_1 t)$. The remaining terms are relatively so small that they can be ignored. Thus, for large t, $X_0 \exp(b_1 t)$ can be cancelled from numerator and denominator to obtain b_1/a_{11}. The system returns to X_1^* from any value of X_0. This confirms the result given in (i).

(iii) The third method is numerical integration. This does not require the analytical solution to be known. Rather, the solution is obtained by approximating the differential equation with small, but finite time intervals. This is a very repetitious operation and requires a computer to be practical. The computer cannot handle the infinitesimally small quantity dt, but can approximate it using a small, but finite value of time called Δt, such as 1/100, 1/1000, etc. Replacing dX and dt by ΔX and Δt, respectively, yields

$$\Delta X/\Delta t = X_1(b_1 - a_{11}X_1). \quad (2.20)$$

Multiplying by Δt and adding X_t to both sides yields

$$X_t + \Delta X = X_t + X_t(b_1 - a_{11}X_t)\Delta t, \quad (2.21)$$

where $X_t + \Delta X$ is the new population size after time Δt. By substituting this new value of X into Equation (2.21) the population size at the next time increment can be obtained, and so on. Performing this operation on a computer a final value that approaches X_1^* can be obtained from any initial value of X.

These three methods are of limited use. The first is limited because, for two or more populations, it is generally not obvious by inspection that the populations will return to equilibrium. Direct integration as in (ii) is usually impossible: the solutions are not known. The third method, using a computer effectively to integrate the equations, can be even too tedious for the computer: there are too many possible initial conditions from which to start the numerical integrations. None the less, numerical methods are often the only way to solve some of the problems to be considered. Numerical techniques are not an easy way out even when computer time is unlimited. The very simple numerical method suggested by Equation (2.21) will fail spectacularly for some ecological models. For these models much more complex techniques must be used (Grove, 1966).

The problem

That the equations cannot be solved explicitly creates a problem for which there are several solutions; each solution has advantages and disadvantages. One solution is to simplify the equations. When the population densities remain close to (in the locality of) the equilibrium, the nonlinear equations can be approximated by some simple linear ones that can be solved explicitly. This technique of determining *local stability* has the additional attraction that, from the equations, certain parameters called eigenvalues can be readily obtained. The sign (negative, positive) of the largest eigenvalue indicates whether or not the population densities will return to equilibrium. This, after all, is of more interest than the details of density changes along the way to (or from) the equilibrium. Eigenvalues have been studied intensively and there are useful theorems about their signs.

The disadvantage of this technique is that it tells nothing about model populations subjected to large perturbations where the approximations of linearity no longer hold. For large perturbations, local stability may fail to capture the essential features of the system's dynamics. Some physical analogies are useful. A stable system is like a ball in a bowl which returns to the lowest part of the bowl (equilibrium) when moved. The region from which the ball would return (called the domain of attraction) may be very small (Fig. 2.1(a)) and surrounded by a very large region from which the ball would not return. Conversely, the equilibrium point may be unstable but surrounded by a large region from which the ball cannot escape (Fig. 2.1(b)).

There are conditions which, if met, ensure that the species' densities return to equilibrium from any combination of densities. This is the property of

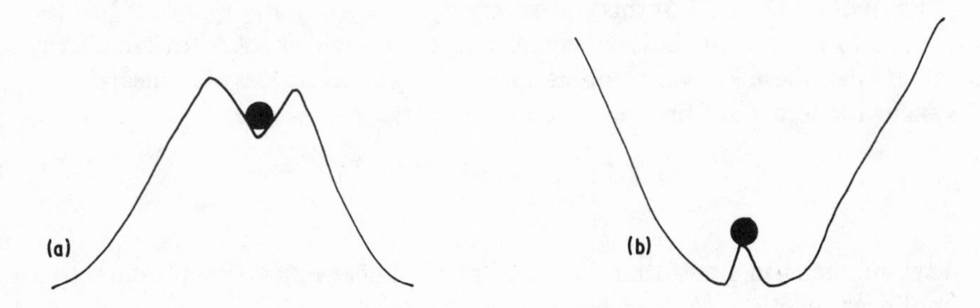

Figure 2.1 Physical analogies of the problems that can arise with local stability analyses. Both are balls (black circles) resting on landscapes whose cross sections are sketched. (a) A locally stable system with a small domain of attraction surrounded by a large region from which the ball will not return. (b) A locally unstable system surrounded by a large region that will retain the ball.

global stability. The model in Equation (2.17) has this property, but more complex models often do not. The purpose of the brief disscussion of global stability in Chapter 3 is sufficient to show that, even when these conditions are met, it may be very difficult to determine them. Worse still, useful mathematical tools to search for global stability are scarce. However, other criteria involving large perturbations are tractable and will be developed in Chapter 3.

2.3.2 Local stability I: differential equations

(a) Step one: linearizing the equations

The importance of considering small perturbations is that, provided populations remain sufficiently close to the equilibrium, the complicated equations can be approximated with linear ones. Linear equations *can* be solved. In Equation (2.17), $\dot{X}_1$ depends on both X_1 and X_1^2; but, at any point, the relationship between $\dot{X}_1$ and X_1 can be approximated by a straight line. In general, a function at a point can be approximated by a straight line, a quadratic equation, a cubic equation, and so on. The more terms in the equation chosen, the better it fits the original function over large ranges of the variable. It is Taylor's expansion that enables the terms to be found in these linear, quadratic, cubic, etc., equations. (A statement of the theorem and examples of its use are given in Appendix 2A.) Application of this expansion to models with more than one species is only slightly more complicated. For two species, a plane is employed to approximate the nonlinear surface. The geometry of this approximation is also discussed in Appendix 2A. For more than two species, a linear surface, called a hyperplane, is employed. This may defy imagination but its useful property, linearity, is not altered. As an example, consider the two-species predator–prey model of Equation (2.7). First, obtain the equilibrium densities. Setting each $\dot{X}_i$ to zero and ignoring the trivial solutions yields linear equations which can be solved simultaneously (see Equation (2.11)). For this two-species system, $X_1^* = b_2/a_{21}$ and $X_2^* = (b_1 - a_{11}X_1^*)/a_{12}$. In the subsequent discussion I shall use X_i^* for simplicity, rather than these known combinations of parameter values. To linearize the system the terms are first rewritten without the parentheses:

$$\dot{X}_1 = b_1X_1 - a_{11}X_1^2 - a_{12}X_1X_2 \tag{2.22a}$$

$$\dot{X}_2 = -b_2X_2 + a_{21}X_1X_2. \tag{2.22b}$$

Taylor's theorem states that the terms of the linear approximation $g(z)$ to a nonlinear function $f(z)$ can be obtained from

$$g(z) = f(z^*) + (z - z^*)f'(z^*), \tag{2.23}$$

where $f(z^*)$ is the value of the nonlinear function at z^* and $f'(z^*)$ is the derivative of the function, evaluated at z^*. As discussed in Appendix 2A, this is only part of Taylor's expansion. There are terms involving $(z-z^*)^2$, $(z-z^*)^3$, etc. These can be safely ignored only if $(z - z^*)$ is small, that is, if values of z are taken near z^*. Finally, expression (2.23) refers to only one variable and to it must be added similar linear expressions to find how the function varies with the other variable(s). To illustrate this, apply Equation (2.23) to Equation (2.22a) for $\dot{X}_1$, noting that a function $g(\dot{X}_1)$ is to be obtained from the more complicated function for $\dot{X}_1$, that is, Equation (2.22a) evaluated at equilibrium, X_1^*:

$$g(\dot{X}_1) = f(X_1^*) + (X_1 - X_1^*)(b_1 - 2a_{11}X_1^* - a_{12}X_2^*). \qquad (2.24)$$

Now, at equilibrium, $\dot{X}_1 = 0$ by definition, and therefore either $b_1 - a_{11}X_1^* - a_{12}X_2^*$ or X_1 must be zero. If the trivial equilibrium is ignored, the terms $b_1 - a_{11}X_1^* - a_{12}X_2^*$ can be taken from (2.24) without affecting its value. The $f(X_1^*)$ is Equation (2.22a) evaluated at equilibrium, which is zero by definition. Thus

$$\dot{X}_1 = (X_1 - X_1^*)(-a_{11}X_1^*). \qquad (2.25)$$

In a similar manner, the variation of $\dot{X}_1$ with X_2 can be obtained:

$$\dot{X}_1 = (X_2 - X_2^*)(-a_{12}X_1^*). \qquad (2.26)$$

These two equations can be added to get the equation for a plane:

$$\dot{X}_1 = (X_1 - X_1^*)(-a_{11}X_1^*) + (X_2 - X_2^*)(-a_{12}X_1^*). \qquad (2.27)$$

Finally, by the same process, the comparable expression for $\dot{X}_2$ can be found:

$$\dot{X}_2 = (X_1 - X_1^*)(a_{21}X_2^*). \qquad (2.28)$$

Note that $\dot{X}_2$ does not depend on X_2. To simplify, let $X_i - X_i^*$ be called x_i, which are the displacements from equilibrium, in short, the perturbations. Note that because X_i and x_i differ by only a constant (X_i^*) that $\dot{X}_i = \dot{x}_i$. Again for simplicity, the constants $a_{ij}X_i^*$ are replaced by new ones of the form c_{ij}. For convenience, absorb the negative signs into the c_{ij} when Equations (2.27) and (2.28) become

$$\begin{aligned} \dot{X}_1 &= \dot{x}_1 = c_{11}x_1 + c_{12}x_2 \\ \dot{X}_2 &= \dot{x}_2 = c_{21}x_1. \end{aligned} \qquad (2.29)$$

Now the problem of solving Equation (2.29) must be addressed. But the question of stability can be rephrased: do the perturbations, x_i, tend to zero as time increases?

(b) Step two: solving the linear equations

Although the Lotka–Volterra equations (e.g. Equation 2.7) cannot be solved, solutions can be found for (2.29). For Equations (2.29), there is an implied c_{22} term (the effect of X_2 on its own growth rate) that is zero. I shall retain the term in the discussion in order to display, more generally, where the various terms arise. The solutions to (2.29) give the perturbations as explicit functions of time (t):

$$x_1 = p_1 m_{11} \exp(\lambda_1 t) + p_2 m_{12} \exp(\lambda_2 t)$$
$$x_2 = p_1 m_{21} \exp(\lambda_1 t) + p_2 m_{22} \exp(\lambda_2 t). \tag{2.30}$$

The p, λ and m are all constants and will be discussed presently. This solution to Equation (2.29) will not be derived. But, as is often the case with integration, it is easy to confirm that it is the correct answer by differentiating it to get the original equations. This I shall do. But first, note that the signs of the λ (the *eigenvalues*) are critical: if both are negative the perturbations (x) will die out as time increases. If either is positive, the perturbations will become larger and larger, which means that the system is unstable because the population densities will move away from the equilibrium. In short, only the *sign* of the largest eigenvalue is needed to determine whether or not the system is stable. Simply, the other constants are unimportant.

Now, I show that (i) the p are dependent on the initial population densities (those following the perturbation); (ii) the m are determined from the constants, c; and (iii) the λ are also uniquely determined by the c. Points (ii) and (iii) emerge in the process of demonstrating that Equations (2.30) are, indeed, solutions to Equation (2.29). First, however, consider the p. Immediately following the perturbations, when $t = 0$, the x have the values given by the direction and magnitude of the perturbations; call these $x_{i(0)}$. Equations (2.30) become

$$\begin{aligned} x_{1(0)} &= p_1 m_{11} + p_2 m_{12} \\ x_{2(0)} &= p_1 m_{21} + p_2 m_{22}. \end{aligned} \tag{2.31}$$

Let me anticipate a result about the m which I shall demonstrate presently: they are known functions of the parameters c. Thus, given a knowledge of the initial conditions ($x_{i(0)}$), Equations (2.31) can be solved for p.

Next, differentiate Equation (2.30) to obtain $\dot{x}_i$ and substitute $\dot{x}_i$ and x_i into Equation (2.29) to see if (2.30) are, indeed, solutions to (2.29). As will become clear, this will also produce expressions that allow the calculation of the m and λ. First consider x_1:

$$\begin{aligned} &\lambda_1 p_1 m_{11} \exp(\lambda_1 t) + \lambda_2 p_2 m_{12} \exp(\lambda_2 t) \\ &\quad = c_{11}[p_1 m_{11} \exp(\lambda_1 t) + p_2 m_{12} \exp(\lambda_2 t)] \\ &\quad + c_{12}[p_1 m_{21} \exp(\lambda_1 t) + p_2 m_{22} \exp(\lambda_2 t)]. \end{aligned} \tag{2.32}$$

Now, the terms $\exp(\lambda_1 t)$ and $\exp(\lambda_2 t)$ are continuously changing with time. So Equation (2.32) can be true only if the terms in $\exp(\lambda_1 t)$ *and* the terms in $\exp(\lambda_2 t)$ on both sides of the equation are equal. That is, if

$$\lambda_1 p_1 m_{11} \exp(\lambda_1 t) = c_{11} p_1 m_{11} \exp(\lambda_1 t) + c_{12} p_1 m_{21} \exp(\lambda_1 t)$$

and

$$\lambda_2 p_2 m_{12} \exp(\lambda_2 t) = c_{11} p_2 m_{12} \exp(\lambda_2 t) + c_{12} p_2 m_{22} \exp(\lambda_2 t). \quad (2.33)$$

Cancellation of terms in p and $\exp(\lambda_i t)$ from both these equations yields

$$\lambda_1 m_{11} = c_{11} m_{11} + c_{12} m_{21} \quad (2.34a)$$

$$\lambda_2 m_{12} = c_{11} m_{12} + c_{12} m_{22}. \quad (2.34b)$$

The answer to whether (2.30) are solutions to (2.29) is a little clearer: it is 'yes' provided constants (m, λ) can be found that fit Equations (2.34). Applying the processes in Equations (2.32)–(2.34) to the equation for x_2 produces the additional equations

$$\lambda_1 m_{21} = c_{21} m_{11} + c_{22} m_{21} \quad (2.35a)$$

$$\lambda_2 m_{22} = c_{21} m_{12} + c_{22} m_{22}. \quad (2.35b)$$

Consider Equations (2.34a) and (2.35a) compared to (2.34b) and (2.35b). They involve the same terms in c but two different terms in λ and $m(m_{11}, m_{21}$ in the 'a' set and m_{22}, m_{12} in the 'b' set). This suggests that there are two solutions to the four equations. Writing λ for λ_1 or λ_2 and m_1 and m_2 for m_{11}, m_{21} and m_{12}, m_{22}, respectively, and making the substitutions yields

$$\lambda m_1 = c_{11} m_1 + c_{12} m_2$$

$$\lambda m_2 = c_{21} m_1 + c_{22} m_2. \quad (2.36)$$

If expressions for m_1/m_2 are obtained from both of these equations and set equal to each other, then

$$-c_{12}/(c_{11} - \lambda) = (c_{22} - \lambda)/-c_{21}. \quad (2.37)$$

Or

$$\lambda^2 - \lambda(c_{11} + c_{22}) + c_{11}c_{22} - c_{12}c_{21} = 0. \quad (2.38)$$

This is a quadratic equation with, indeed, two possible values for λ:

$$\lambda = \{(c_{11} + c_{22}) \pm [(c_{11} + c_{22})^2 - 4(c_{11}c_{22} - c_{12}c_{21})]^{1/2}\}/2 \quad (2.39)$$

Substituting each of these values of λ into Equation (2.35) gives the two sets of values of m. An example of stages (2.27)–(2.39) is shown in Appendix 2B.

This result is an interesting one. The expression for the eigenvalues, λ, depends only on the parameters, the terms in c, and not on the initial perturbations. And the sign of the eigenvalues determines whether or not the system returns to equilibrium from *any* perturbation that is small enough for the linearized equations to be valid.

(c) Eigenvalues

Although the above derivation may seem tedious, the result is elegant. Equations (2.30) have a number of terms, but only one, the largest eigenvalue, is required to decide whether the perturbations will or will not become zero as time increases. Given the b_i and a_{ij} which define the model, the terms in c and, from Equation (2.39), the eigenvalues can be calculated. Although the exact changes in density would require calculation of the p and m, these are not needed to decide whether the perturbations die out or become larger. The work is simplified as a consequence. Now, there are several possibilities that arise in solving the quadratic Equation (2.39):

(i) If either or both of the eigenvalues are positive, the perturbations will continue to grow, that is, the system will move away from equilibrium.

(ii) If both of the eigenvalues are negative the perturbations will die out. The system is deemed stable because the species densities return to equilibrium.

(iii) Suppose the solution to (2.38) involves complex numbers, that is, the term within the square root of (2.39) is negative. The eigenvalues become $r \pm \mathrm{i}q$, where i is the square root of minus one (and is not to be confused with the ith species). This circumstance has a particular meaning. Euler's formula (derived in most introductory calculus texts, e.g., Flanders, Korfhage & Price, 1970, p. 722) states that

$$\exp(r+\mathrm{i}q) = \exp(r)(\cos q + \mathrm{i} \sin q). \tag{2.40}$$

Substituting this into the appropriate term in the solution yields

$$\exp[t(r+\mathrm{i}q)] = \exp(rt)(\cos qt + \mathrm{i} \sin qt). \tag{2.41}$$

First, note that the solution will still approach zero (the system will be stable) only if r (which is called the real part of the eigenvalue, $\mathrm{Re}(\lambda)$) is less than zero. Second, the solution contains trigonometric functions (sines, cosines) and they ensure that the approach to equilibrium is cyclical. Now the presence of i in the solution may seem a problem, for population densities can hardly be imaginary. The solution, however, also involves p and m. These are obtained by substituting the values of λ into Equations (2.31) and (2.36), respectively, which reveals that these sets of terms are also complex. The resulting solution can be written in full and the terms collected. The algebra is too lengthy to repeat here, but all the imaginary terms cancel. Leighton (1970, pp. 168–9) presents a more elegant derivation of this result, which avoids the extensive use of the algebra just suggested.

The possibilities (i)–(iii) are illustrated in Fig. 2.2. The left-hand diagrams contain plots of predator and prey densities independently against time. The right-hand diagrams contain equally informative and visually much simpler plots of successive pairs of predator and prey densities. Here, the times for the densities to return to the two-species equilibrium are not apparent, but whether the system returns to equilibrium is apparent.

(d) Return times

The largest eigenvalue determines not only whether the system will return to equilibrium but, if it does return, how fast it will do so. Suppose that for a two-

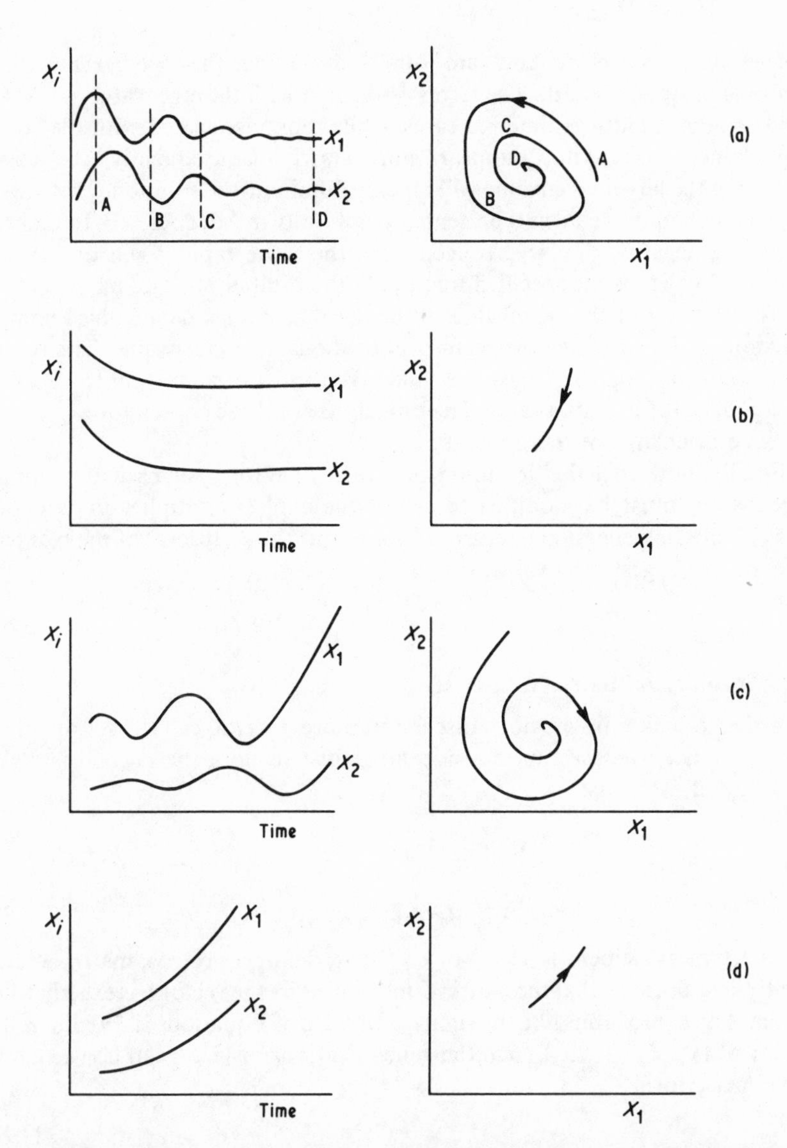

Figure 2.2 Dynamics of a two-species system. In the figures on the left, each species population density (X_1, X_2) is plotted against time. In the figures on the right, one species density is plotted against the corresponding density of the other species at each point in time. Arrows indicate the progression of these points in time. For example, four points in time (A, B, C, D) are shown in the top two figures. The four cases shown are: (a) a stable system with densities oscillating as they return to the two-species equilibrium (both eigenvalues are negative but complex); (b) a stable system with monotonic return to equilibrium (both eigenvalues negative and not complex); (c) an unstable system (one or both eigenvalues positive and complex); and (d) an unstable system (one or both eigenvalues positive but not complex).

species system the eigenvalues are -0.1 and -0.9 (as they are in the worked example in Appendix 2B). The terms associated with the eigenvalue -0.9 will tend to zero rapidly as time increases, while those associated with the larger eigenvalue, -0.1, will disappear more slowly. Consequently, the terms involving the larger eigenvalue will quickly dominate the solution. This idea is explained in more detail and presented graphically in Appendix 2B. In general, the more negative the largest eigenvalue, the more rapidly will the perturbations decay to some specified fraction of their initial values. This suggests a useful function of the eigenvalues to be the *return time*, defined by Pimm & Lawton (1977) as minus the reciprocal of the largest eigenvalue. The return time measures approximately the time taken for a perturbation to decay to 1/e ($\simeq 37\%$) of its initial value. The time units are those chosen for expressing the rate of change of the populations.

Finally, note that the definition of return time for systems with complex eigenvalues must be modified to reflect the exp(rt) multiplier in Equation (2.41). Thus, in general, the return time is minus the reciprocal of the real part (Re) of the largest eigenvalue, that is

$$\text{Return time} = -1/\text{Re}(\lambda_{\max}), \qquad \text{for } \lambda_{\max} < 0. \tag{2.42}$$

(e) Stability in matrix terms

In order to make the previous discussion more general, call the set of values $x_1, x_2, \ldots, x_n$ a *vector* **x** and consider an entity $\boldsymbol{C}$, called a matrix, consisting of an array of elements:

$$\begin{matrix} a_{11}X_1^* & a_{12}X_1^* & \ldots & a_{1n}X_1^* \\ \vdots & & & \vdots \\ a_{n1}X_n^* & a_{n2}X_n^* & \ldots & a_{nn}X_n^*. \end{matrix} \tag{2.43}$$

Thus a typical element is $a_{ij}X_i^*$ or c_{ij}. Next, define a process, matrix–vector multiplication, such that the matrix $\boldsymbol{C}$ multiplied by the vector **x** is exactly what is meant by expressions like the right-hand sides of Equations (2.29). If $\dot{\mathbf{x}}$ is the vector of $(\dot{x}_1, \dot{x}_2, \ldots, \dot{x}_n)$, then the generalizations of (2.29) can be written in the concise form

$$\dot{\mathbf{x}} = \boldsymbol{C}\mathbf{x}. \tag{2.44}$$

Nothing new is involved here; Equation (2.44) is just another notation. Equations (2.36) can also be written in this concise form as a vector **m** multiplied by a constant (the eigenvalue) on the left-hand side of the expression and by a matrix $\boldsymbol{C}$ on the right-hand side, so that

$$\lambda\mathbf{m} = \boldsymbol{C}\mathbf{m}. \tag{2.45}$$

Eigenvalues are defined by this expression, and the m with which they are associated are called eigenvectors. Again this does not create anything new; the eigenvalues must still be calculated using some algebraic manipulations equivalent to Equations (2.37)–(2.39) on the *elements* of the matrix $\boldsymbol{C}$. This matrix is sufficiently important to have the name *Jacobian matrix* (defined in Appendix 2C). Jacobian matrices for model systems other than Lotka–Volterra will not have elements of the same form as in Equation (2.43).

Biologically, the elements of the Jacobian matrix are, in all cases, the effects of one species' density upon another's growth rate when all the species are at their equilibrium densities. This result is particularly important. The signs and the approximate magnitudes of these matrices can be estimated with more certainty than the dynamics of the system over large ranges of species densities. In short, more is known about the Jacobian matrix than the system it approximates. Because only the matrix elements are required to determine local stability, it is not necessary to develop a model and linearize it, as presented in the arguments above. Of course, the discussion is valuable as an example of how the matrix coefficients are derived, which is the reason why the material was included. But it should be stressed that, in subsequent chapters, the assumptions are based on the matrix elements directly, combined with the assumption that populations remain close enough to equilibrium that linearizations are valid.

(f) Results about the eigenvalues of many-species systems

In Equation (2.39) I obtained the eigenvalues as the roots of a quadratic equation. For a three-species system I would have to solve a cubic equation, and for four species a quartic equation, and so on. There are as many eigenvalues as equations (species) in the system. Beyond a quartic equation, explicit solutions cannot, in general, be found for the eigenvalues. Indeed, the solution for a quartic is extremely complicated.

It may seem rather pointless developing all this material only to find that the eigenvalues cannot be determined. However, the eigenvalues of a particular matrix can be found on a computer very quickly and to a high degree of accuracy using numerical methods. More important, there are some very useful theorems about the eigenvalues of matrices and, particularly, about the largest eigenvalue of a matrix. I shall present several of these throughout the text. Next, I shall consider conditions that ensure that a matrix has negative eigenvalues regardless of the magnitude of the coefficients in the Jacobian matrix.

Qualitative stability

The signs of the interactions in the Jacobian matrix reflect the trophic relationships between species (e.g. $c_{ij} < 0$, $c_{ji} > 0$, indicate that X_i is the prey of X_j). Indeed, the sign structure of the matrix of the c_{ij} (or a_{ij}) and the food web are simply alternative representations of each other. Often, the exact magnitudes of the c_{ij} are not known, so a particularly useful concept is that of *qualitative stability*. Suppose the magnitudes, but not the signs, of every element in the Jacobian matrix are changed. If every new system produced in this way has strictly negative eigenvalues the system is considered to be qualitatively stable (Jeffries, 1974). The conditions on the signs of the matrix elements that are sufficient to ensure qualitative stability are:

(i) $c_{ii} \leqslant 0$ for all i.
(ii) $c_{ii} < 0$ for some i (which particular i depends on the model, see later).
(iii) $c_{ij}c_{ji} < 0$ for all $i \neq j$.
(iv) $c_{ij}c_{jk} \ldots c_{qr}c_{ri} = 0$ for any sequence of three or more distinct indices $i, j, k, \ldots, q, r$.
(v) The determinant of the matrix $\boldsymbol{C}$ is not zero. (The term determinant will be discussed shortly.)

The proof of these conditions and their origin are given in Jeffries (1974). The first condition comes from an important result that the sum of the eigenvalues equals the sum of the diagonal elements of the matrix. (This is demonstrated, though not proved, in the worked example in Appendix 2B.) If even one diagonal element is positive, then it may be sufficiently large that one or more of the eigenvalues must also be positive. Other things being equal, the more negative terms on the diagonal of the Jacobian matrix, the more likely all the eigenvalues will be negative and the system stable.

Some of these conditions have ecological implications that are obvious, while others are less so. Condition (iii) prevents models with direct interspecific competition ($c_{ij} < 0$, $c_{ji} < 0$) or mutualism ($c_{ij} > 0$, $c_{ji} > 0$) from being qualitatively stable. Conditions (i) and (ii) mean that there must be at least one species in the system that is limited by some resource from outside the system. This is not contentious, but it has been argued that all the diagonal elements are less than zero, and this is far from certain. I shall return to this topic in Chapter 4. If only some of the c_{ij} are negative, it becomes crucial to determine exactly which species are the 'some' in condition (ii). This approach to the problem will also clarify condition (iv). I discuss condition (v) last.

The development of conditions (ii) and (iv) entails the use of directed graphs, called digraphs for short. A digraph consists of points for each of the n species and directed lines (up to n^2 in all) indicating whether or not a species affects another species' growth rate (or its own). A simple digraph of a two-species predator–prey system is given in Fig. 2.3(a).

Jeffries defines a *p cycle* as a set of p distinct points with the property that a circuit may be traced among the p points by directed lines. Each p cycle involves precisely p lines – figures of eight are not considered cycles. In Fig. 2.3(a) there is a 1-cycle (X_1 to X_1) and a 2-cycle (X_1 to X_2 to X_1). Condition (iv) may be restated: there can be no p cycles with $p > 2$. The web in Fig. 2.3(b) is not qualitatively stable because X_3 (carnivore) feeds on X_2 (herbivore) as well as X_1 (plant). There are two 3-cycles (X_1 to X_2 to X_3 to X_1 and X_1 to X_3 to X_2 to X_1). Without the carnivore–plant interaction, the system would lack a 3-cycle and would, indeed, be qualitatively stable.

To develop condition (ii), I must describe what Jeffries calls a 'predation community'. Jeffries writes:

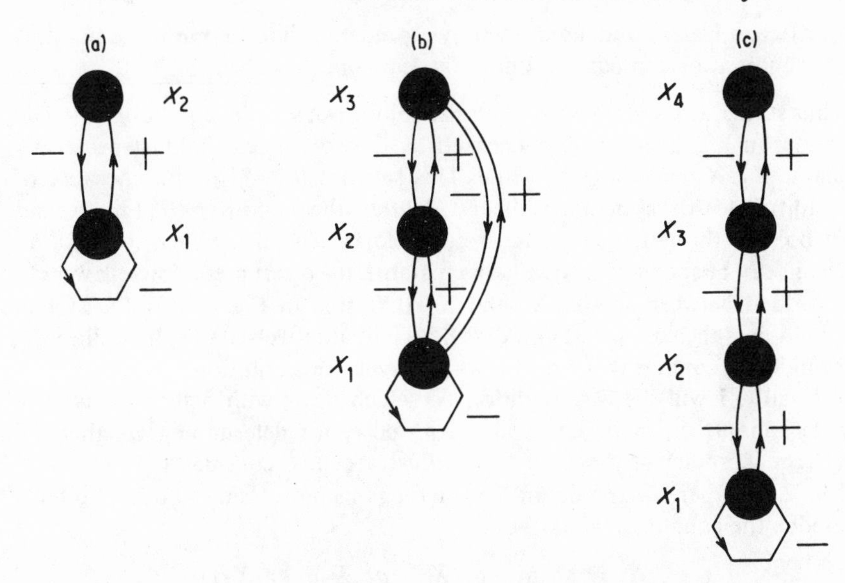

Figure 2.3 Digraphs of (a) two, (b) three and (c) four trophic level systems whose qualitative stability is discussed in the text. Circles represent species, lines the interactions between species. A species below is a prey to the species above it on the page. Signs indicate the effect of the density of a species on the growth rate of the species with which it interacts.

> '. . . if two species are involved in a 2-cycle and, if the 2-cycle involves one positive line and one negative line, then the species may be regarded as predator and prey . . .. The species are said to be related by a 'predation link'. Associate with a fixed species all the other species, if any, to which it is related by predation links. Then, associate with these species all the additional species related by predation links and so on. The maximal set of all such species related to the first species is called the 'predation community' containing the first species. For the sake of completeness, a species not connected by a predation link to any other species is also called a predation community, albeit trivially so.'

Jeffries continues by defining a 'colour test' that supposes the points (that represent the species) to be one of two colours.

> 'A predation community passes the colour test provided each point in the associated digraph may be coloured black or white with the result that:
> (a) each self-regulating point ($c_{ii} < 0$) is black;
> (b) there is at least one white point;
> (c) each white point is connected by a predation link to at least one other white point;

(d) each black point, connected by a predation link to one white point, is connected by a predation link to at least one other white point.'

Thus stated, any system which *fails* the colour test satisfies condition (ii). The system in Fig. 2.3(c) may be checked by the colour test: colour species X_1 black; X_2, X_3, and X_4 are white. This satisfies (b) and (c) but, because of condition (d), the system fails the test. All the other conditions (i), (iii), (iv) and (v) hold, so the system is qualitatively stable. In general, simple n-species food chains can be shown to be qualitatively stable. If a direct predation link were to be added between species X_1 and X_3 (like that of Fig. 2.3(b)), the system would pass the colour test and thus not be qualitatively stable. In addition, it would also contain two 3-cycles which invalidate condition (iv).

Finally, I will discuss condition (v) which deals with determinants. The determinant of a matrix is a useful property for determining whether the system is sufficiently specified. To illustrate this, consider the system in Fig. 2.4(a) where two predators feed on the same prey. Using a Lotka–Volterra model, the equations would be

$$\begin{aligned} \dot{X}_1 &= X_1(b_1 - a_{11}X_1 - a_{12}X_2 - a_{13}X_3) \\ \dot{X}_2 &= X_2(-b_2 + a_{21}X_1) \\ \dot{X}_3 &= X_3(-b_3 + a_{31}X_1). \end{aligned} \tag{2.46}$$

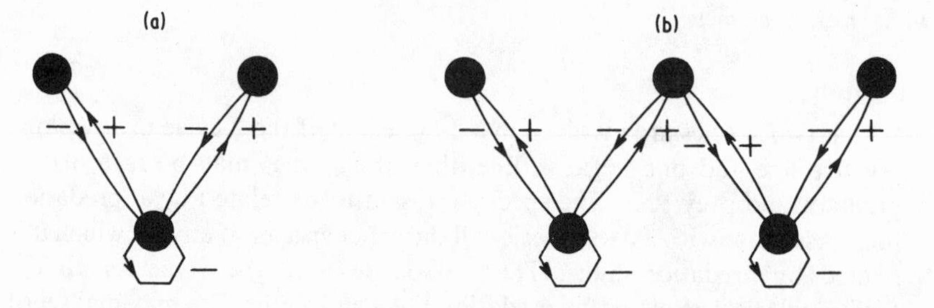

Figure 2.4 Digraphs of singular systems discussed in the text. Sign conventions as in Fig. 2.3.

To obtain the equilibrium densities, the recipe involves setting each $\dot{X}_i$ to zero and solving the expressions within the parentheses. For the system (2.46), this is impossible. There are two equations for X_1 which cannot be solved simultaneously except by extraordinary biological coincidence. Moreover, there is only one equation for the two variables X_2, X_3. The determinant of the Jacobian matrix of this system is zero, and such a system is said to be *singular*. A singular system can arise in several ways but, commonly, it involves any m species of predator feeding on p species of prey, where p is less than m. Another example is shown in Fig. 2.4(b) which illustrates three species of predator feeding on two species of prey. The condition is really a mathematical statement of Gause's hypothesis (see Lack, 1971) that there cannot be more than two species exploiting exactly the same niche. Now, two predators often do exploit almost exactly the same prey species in the real world. They usually do so, however, by exploiting the prey in diffierent habitats, geographical areas or times of day or year (Schoener, 1974). Such predators belong to effectively different food webs.

2.3.3 Local stability II: difference equations

The recipe for determining the stability of pairs of equations like (2.13) and (2.14) is very similar to the process outlined above for differential equations. First, solve for the equilibrium by noting that this occurs when $X_{t+1} = X_t$, that is, there is no change in the population size. Then, using Taylor's theorem, expand the nonlinear equations about the equilibrium to obtain the linear equations. For a two-species system these will be of the form

$$\begin{aligned} x_{1,t+1} &= c_{11}x_{1,t} + c_{12}x_{2,t} \\ x_{2,t+1} &= c_{21}x_{1,t} + c_{22}x_{2,t}, \end{aligned} \tag{2.47}$$

where the c_{ij} are constants and the x_i are the differences between the population levels and the equilibrium densities. The form of the solution is

$$\begin{aligned} x_{1,t} &= m_{11}\lambda_1^t + m_{12}\lambda_2^t \\ x_{2,t} &= m_{21}\lambda_1^t + m_{22}\lambda_2^t. \end{aligned} \tag{2.48}$$

As before, I shall not derive the solution but will show that it works, provided certain relationships hold. Substitute Equations (2.48) into (2.47). For x_1, this gives

$$\begin{aligned} m_{11}\lambda_1^{t+1} + m_{12}\lambda_2^{t+1} &= c_{11}(m_{11}\lambda_1^t + m_{12}\lambda_2^t) \\ &\quad + c_{12}(m_{12}\lambda_1^t + m_{22}\lambda_2^t). \end{aligned} \tag{2.49}$$

Because λ^t is constantly changing, as $t = 0, 1, 2, 3, \ldots$ etc., this equality can be true only if the terms in λ_1 and λ_2 are equal. That is

$$m_{11}\lambda_1^{t+1} = c_{11}m_{11}\lambda_1^t + c_{12}m_{21}\lambda_1^t$$

and

$$m_{12}\lambda_2^{t+1} = c_{11}m_{12}\lambda_2^t + c_{12}m_{22}\lambda_2^t. \tag{2.50}$$

Cancellation of terms in λ_i^t from both these equations gives

$$\begin{aligned} m_{11}\lambda_1 &= c_{11}m_{11} + c_{12}m_{21} \\ m_{12}\lambda_2 &= c_{11}m_{12} + c_{12}m_{22}. \end{aligned} \tag{2.51}$$

A similar process for x_2 yields

$$\begin{aligned} m_{21}\lambda_1 &= c_{21}m_{11} + c_{22}m_{21} \\ m_{22}\lambda_2 &= c_{21}m_{12} + c_{22}m_{22}. \end{aligned} \tag{2.52}$$

Equations (2.51) and (2.52) are identical to (2.34) and (2.35) and the subsequent argument, including its generalization to n species, is identical. The λ_i are the eigenvalues. But note that if the system is to be stable, the solution (Equation (2.48)) places different limits on the eigenvalues than the comparable equations for differential equations (2.30). For the perturbations ($x_{i,t}$) to become zero as time increases all the λ_i must be less than +1. A fraction of unity clearly becomes smaller as it is raised to higher and higher powers as time increases 1, 2, . . ., ∞. The eigenvalues must also be greater than −1, otherwise the solution will move away from equilibrium, oscillating (from positive values to negative values) as it goes. As before, the eigenvalues can have imaginary parts and, as before, if they do, the solution will also oscillate. Only if the eigenvalues are all real and lie between zero and one will the populations approach their two-species equilibrium monotonically. Stability places limits on the eigenvalues that affect the magnitude of both their real (Re) and imaginary (Im) parts. The exact condition is

$$[\mathrm{Re}(\lambda_i)]^2 + [\mathrm{Im}(\lambda_i)]^2 < 1, \quad \text{for all eigenvalues.} \tag{2.53}$$

The result is derived by May (1973b). It is easy to visualize this result: the eigenvalues must lie within a circle of radius unity, centred at the origin of the complex plane (the y axis is the magnitude of the imaginary part and the x axis the magnitude of the real part of the eigenvalue).

2.4 SUMMARY

2.4.1 Models

Three formats are presented which model the interactions between populations. The first and second are in differential equation form – the changes in population densities are continuous. The third is in difference equation form – the changes in population densities occur at discrete points in time. The former are appropriate to populations with overlapping generations in relatively aseasonal environments. The difference equations are more appropriate to species, like insects, with discrete generations and which spend a portion of their year in an effectively non-interactive state, such as diapause. The difference between the two differential forms lies in the effect which the predators exert on the growth rate of their prey. In the first format, which I call Lotka–Volterra, predators exert a negative effect on the growth rate of their prey. In the second format, which I call donor-control, they do not: the predators take only those prey which would die (or have died) from other causes. The choice of these three formats is deliberate: they differ considerably in their consequences to the stability of food web structures.

2.4.2 Stability

Systems are deemed stable when all species densities return to equilibrium following a perturbation. Determining the stability of multispecies models is not easy. It is necessary to solve the difference and differential equations, that is, to know the population densities as functions of time. These solutions cannot be found except in very special and unrealistically simple cases. It is possible, however, to produce linear approximations to the nonlinear equations for population densities near (local to) the multispecies equilibrium. The linear equations can be solved, but only at a price: the solutions are valid only for a small region near the equilibrium. This is the condition of local stability. Other definitions of stability are discussed in the next chapter.

The recipe for determining local stability has three steps:

First, the equilibrium densities, X_i^*, must be obtained. These are, by definition, where the populations remain at constant levels, that is, where all $\dot{X}_i = 0$ (for differential equation models) or, equivalently, where $X_{i,t+1} = X_{i,t}$ (for difference equation models). This gives a series of equations that can be solved simultaneously in all except some special cases. The equilibrium densities are obtained as functions of the various parameters in the models.

Second, the equations must be linearized about this equilibrium. To do this Taylor's theorem is employed – a commonsensical notion that a complicated function can be approximated successively by fitting a straight line, a quadratic curve, and so on. The linear equations give the rate of change of the difference between each species' density and its equilibrium (the perturbation) as a function of the perturbations and a set of constants, called the Jacobian matrix. These constants are derived from the parameters of the model and are the effects of each species upon the growth rate of all the other species, evaluated at equilibrium. In an n-species model there are, therefore, n^2 constants. In the Lotka–Volterra models, each constant is the product of a species' equilibrium density and the *per capita* effect of one species upon the growth rate of the other.

Third, special numbers (n in all) called eigenvalues are obtained (or can at least be closely approximated) by algebraic manipulations on the Jacobian matrix. The eigenvalues are critical. The largest eigenvalue contains all the information to characterize whether or not *all* the species densities will return to equilibrium from *any* perturbation small enough so that the linear equations are valid approximations. Moreover, if the system is stable, the largest eigenvalue gives an approximate measure of how fast the system returns to equilibrium.

There are useful theorems concerning eigenvalues, one such theorem being that of qualitative stability. Here, differential equation systems can be known to be stable irrespective of the magnitude of the interactions between species.

Qualitative stability depends on the signs of the matrix elements and these have a direct correspondence to the structure of the food web which the equations model.

APPENDIX 2A. TAYLOR'S EXPANSION

One dimension

Taylor's theorem states a commonsensical notion that successively better fits to a complicated function can be obtained by using a straight line, then a quadratic curve, a cubic, a quartic, and so on. An example is shown in Fig. 2.5; the function is nonlinear (it is a cubic equation of the form $y = z^3 + pz^2 + qz + r$). Over a small region about the point z^*, a straight line $y = a + bz$ is a good approximation, though it fails spectacularly as z moves away from z^*. The quadratic function $y = dz^2 + ez + h$ is a better fit. Formally, Taylor's theorem states:

$$f(z) = f(z^*) + (z - z^*)f'(z^*) + (z - z^*)^2 f''(z^*)/2! + (z - z^*)^3 f'''(z^*)/3! + \ldots + (z - z^*)^n f^{(n)}(z^*)/n!. \quad (2A.1)$$

The term $f(z^*)$ is the complicated function evaluated at z^*, f' means this function differentiated once, $f^{(n)}$ means the function differentiated n times, and $n!$ has its usual meaning of $n(n-1)(n-2) \ldots (3)(2)(1)$. The equation gives an appreciation of why it is mandatory that deviations from a point $(z - z^*)$ be

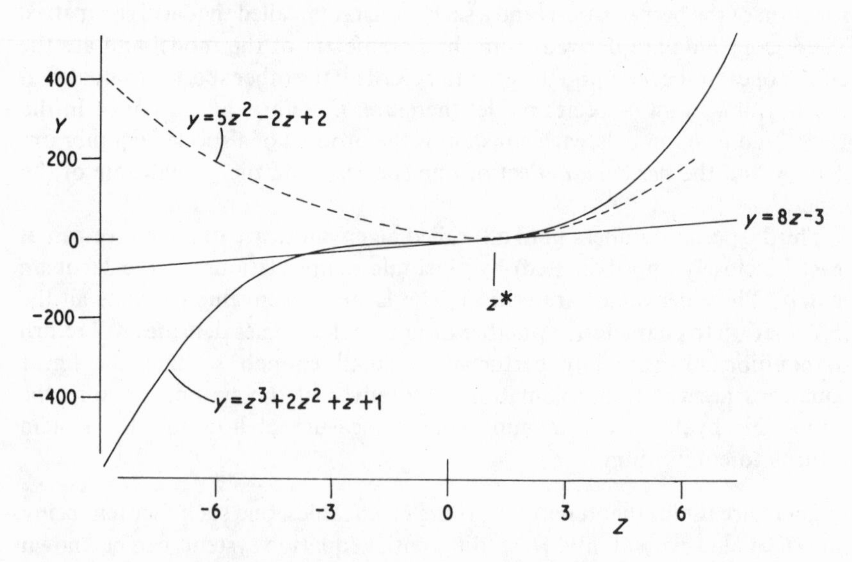

Figure 2.5 A cubic function and its linear and quadratic approximations at $z = 1$ obtained from Taylor's expansion.

small for a linear approximation, for only then will the terms in $(z-z^*)^2$, $(z-z^*)^3$, etc., be small relative to $(z-z^*)$. When the system is close to z^*, the function can be approximated by the function $g(x)$ which contains the first two terms of Equation (2A.1):

$$g(z)=f(z^*)+(z-z^*)f'(z^*). \qquad (2A.2)$$

The example in Fig. 2.5 is the cubic equation

$$y=z^3+2z^2+z+1$$

and

$$z^*=1. \qquad (2A.3)$$

Applying (2A.2) to (2A.3):

$$f(z^*)=1^3+2(1)^2+1+1=5$$

and

$$(z-z^*)f'(z^*)=(z-1)(3(1^2)+4(1)+1).$$

Thus, $g(z)=5+8(z-1)=8z-3$. This is the line shown in the figure. To obtain the quadratic approximation add the next term in Equation (2A.1) to what has already been obtained:

$$(z-z^*)^2f''(z^*)/2=(z-1)^2(6+4)/2=5z^2-10z+5. \qquad (2A.4)$$

Adding (2A.4) to $8z-3$ gives

$$5z^2-2z+2.$$

This is the quadratic function shown in the figure.

More than one dimension

In general, the equations to be approximated involve more than one variable. Consider the following equation with two variables:

$$y=1.09z_1-z_1^2-0.3z_1z_2. \qquad (2A.5)$$

This equation describes a surface, where y is the height of the points above the base and z_1 and z_2 are the axes. A sketch of this function is shown in Fig. 2.6. Suppose this surface is to be approximated by the simplest linear surface, namely, a plane. Recall that a plane is defined by three parameters, an intercept (a) and two slopes (b, c), in the form

$$y=a+bz_1+cz_2. \qquad (2A.6)$$

A sketch of the plane that approximates Equation (2A.5) at the point $z_1=1$ and $z_2=0.3$ is also given in the figure. How are the estimates of a, b and c

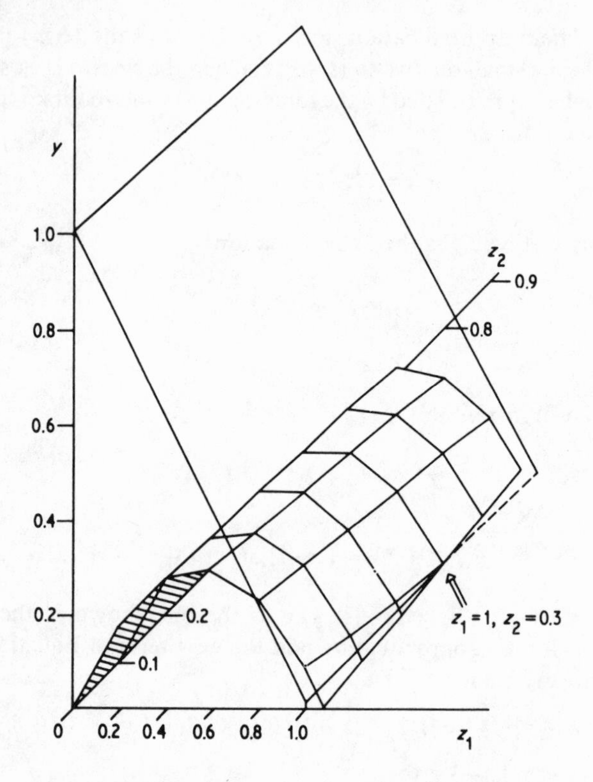

Figure 2.6 A curved surface, $y = 1.09z_1 - z_1^2 - 0.3z_1z_2$ and the plane $y = 1.09 - z_1 - 0.3z_2$ which approximates it at $z_1 = 1$ and $z_2 = 0.3$.

obtained? Proceed by assuming that z_2 is a constant and continue as before:

$$\begin{aligned} g(z_1) &= f(z_1^*) + (z_1 - z_1^*)f'(z_1^*) \\ &= 1.09(1) - (1)^2 - 0.3(1)(z_2) + (z_1 - 1)(1.09 - 2(1) - 0.3z_2). \end{aligned} \quad (2A.7)$$

Now substituting the value of 0.3 for z_2 obtain

$$g(z_1) = 1 - z_1. \quad (2A.8)$$

Now assume that z_1 is constant and obtain the linear function for z_2:

$$\begin{aligned} g(z_2) &= f(z_2^*) - (z_2 - z_2^*)f'(z_2^*) \\ &= 1.09z_1 - z_1^2 - 0.3z_1(0.3) + (z_2 - 0.3)(-0.3z_1). \end{aligned} \quad (2A.9)$$

Substituting the value of $z_1 = 1$ yields

$$g(z_2) = 0.3(0.3 - z_2). \quad (2A.10)$$

Adding (2A.8) and (2A.10) yields the equation for the plane:

$$g(z_1, z_2) = 1.09 - z_1 - 0.3z_2. \quad (2A.11)$$

The three parameters (1.09, -1 and -0.3) are those used in the figure. The two slopes (-1, -0.3) were obtained by differentiating the function with respect to each variable while keeping the other variable constant. Such derivatives are called *partial derivatives.*

While the derivative of a function $y = f(z)$ is usually written dy/dx or, as in this book, f', partial derivatives of a function $y = f(z_1, z_2, \ldots, z_n)$ are written as

$$\frac{\partial y}{\partial z_1}, \frac{\partial y}{\partial z_2}, \ldots, \frac{\partial y}{\partial z_n}. \tag{2A.12}$$

APPENDIX 2B: AN EXAMPLE OF CALCULATING EIGENVALUES

Let

$$\begin{aligned} \dot{X}_1 &= X_1(1.09 - X_1 - 0.3X_2) \\ \dot{X}_2 &= X_2(-1 + X_1). \end{aligned} \tag{2B.1}$$

Obtain the non-trivial equilibrium values, X_1^*, by setting each $\dot{X}_i = 0$ and ignore the possibilities where $X_i = 0$. From the equation for $\dot{X}_2$, $X_1^* = 1$. Substituting this into the other equations yields $X_2^* = 0.3$. To linearize the system use Equations (2.29). (The plane that approximates this equation for $\dot{X}_1$ is derived in detail in the worked example above on Taylor's expansion.) The required terms are the a_{ij} multiplied by the corresponding equilibrium densities, giving in this case

$$\begin{aligned} \dot{x}_1 &= -x_1 - 0.3x_2 \\ \dot{x}_2 &= 0.3x_1. \end{aligned} \tag{2B.2}$$

Using (2.39), the eigenvalues are found to be

$$\begin{aligned} \lambda_i &= \{-1 \pm [1 - 4(0.9)]^{1/2}\}/2 \\ &= -0.9, -0.1. \end{aligned}$$

Both eigenvalues are negative, so the system is stable. The solution to (2B. 2) is

$$\begin{aligned} x_1 &= p_1 m_{11} \exp(-0.9t) + p_2 m_{12} \exp(-0.1t) \\ x_2 &= p_1 m_{21} \exp(-0.9t) + p_2 m_{22} \exp(-0.1t), \end{aligned}$$

where p_i and m_{12} have the meanings discussed in the text. For convenience, suppose that each $p_i m_{ij} = 0.5$. The solution is the sum of two parts, and both are shown in Fig. 2.7. That part involving the smaller eigenvalue (-0.9) decays quickly, leaving the solution approximately equal to the part involving the larger eigenvalue (-0.1). Thus, not only does the larger eigenvalue determine whether the system returns to equilibrium, but also the term containing it quickly comes to dominate the solution. Consequently, the time taken for a

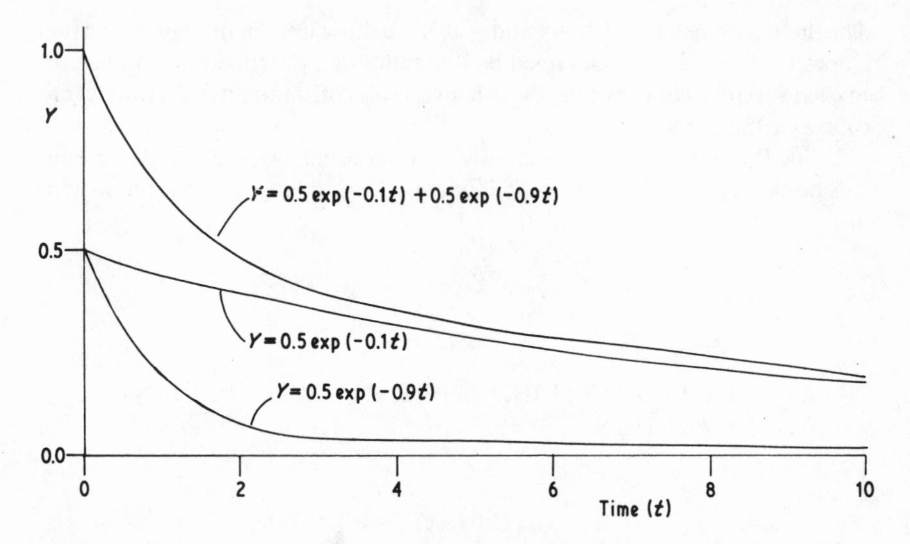

Figure 2.7 Solution to a stable two-species system approximated by Taylor's expansion about the equilibrium (upper curve) and its component parts. Note that the part involving the smaller eigenvalue (-0.9) decays quickly leaving the solution approximately equal to the part involving the larger eigenvalue (-0.1).

perturbation to decay to a specified fraction of its initial value will depend mostly on the largest eigenvalue. The more negative the largest eigenvalue, the shorter this time. This is why the reciprocal of the largest eigenvalue, the 'return time' is a useful parameter.

APPENDIX 2C: JACOBIAN MATRICES

For Lotka–Volterra models it has been shown that stability depends on a set of coefficients called the Jacobian matrix (Equation 2.40). Inspection of the worked example in Appendix 2B and the algebraic processes in the body of the chapter reveals that these coefficients are simple partial derivatives. Each row of n values contains the parameters which describe the slopes (in n dimensions) of the hyperplane which approximates the complicated function of each species' growth rate (written in terms of the other species' densities). Thus, in the two-species example, the growth function for X_1

$$\dot{X}_1 = X_1(1.09 - X_1 - 0.3X_2) \tag{2C.1}$$

was approximated by the plane

$$\dot{X}_1 = 1.09 - X_1 - 0.3X_2 \tag{2C.2}$$

at the equilibrium points $X_1 = 1$ and $X_2 = 0.3$. The elements of the first row of the Jacobian matrix are, therefore, -1 and -0.3. These are the partial derivatives of the function in Equation (2C.1) with respect to X_1 and X_2, respectively, evaluated at the equilibrium.

That the entries of the Jacobian matrix are partial derivatives of the growth equations for each species evaluated at equilibrium is a general result. Consider the system of equations

$$\begin{aligned} \dot{X}_1 &= f_1(X_1, X_2, \ldots, X_n) \\ \dot{X}_2 &= f_2(X_1, X_2, \ldots, X_n) \\ &\vdots \\ \dot{X}_n &= f_n(X_1, X_2, \ldots, X_n) \end{aligned} \tag{2C.3}$$

where the f_i are general functions of the species densities. Linearizing a typical function would yield

$$g(\dot{X}_i) = f_i(X_1^*, X_2^*, \ldots, X_n^*) + (X_1 - X_1^*)\frac{\partial f_i}{\partial X_1} + (X_2 - X_2^*)\frac{\partial f_i}{\partial X_2} + \ldots + (X_n - X_n^*)\frac{\partial f_i}{\partial X_n} \tag{2C.4}$$

where the partial derivatives will be evaluated at the equilibrium densities for each species. Now the functions evaluated at equilibrium ($f_i(X_1^*, X_2^*, \ldots, X_n^*)$) are zero by definition, and, as before, $X_i - X_i^*$ is written as x_i. Finally, $\dot{X}_i^* = \dot{x}_i$, so that the linearized version of (2C.3) becomes

$$\begin{aligned} \dot{x}_1 &= x_1\frac{\partial f_1}{\partial X_1} + \ldots x_n\frac{\partial f_1}{\partial X_n} \\ &\vdots \\ \dot{x}_n &= x_n\frac{\partial f_n}{\partial X_1} + \ldots x_n\frac{\partial f_n}{\partial X_n}. \end{aligned} \tag{2C.5}$$

This can be written in the more compact matrix–vector form

$$\dot{\mathbf{x}} = \mathbf{C}\mathbf{x} \tag{2C.6}$$

which was explained earlier and where the matrix $\mathbf{C}$ can be seen to consist of partial derivatives each of the form $\partial f_i/\partial X_j, i = 1, \ldots, n$ and $j = 1, \ldots, n$.

3 Stability: other definitions

3.1 INTRODUCTION

The purpose of this chapter is to expand the list of meanings of stability to include those less restrictive than that of local stability discussed in the last chapter. Then, by way of summary, I shall attempt to see which of the various combinations of model formats and definitions of stability seem likely to be the most informative in predicting food web structures. The problem with local stability is that if the nonlinear terms are large, there is only a very small region about equilibrium where the linear approximations are accurate. First, consider an example from Goh (1977) which shows that local stability does not imply global stability. A globally stable system is one that returns to equilibrium from any initial conditions, not just those close to the equilibrium. Goh's example is for three competing species, each limited by the other species as well as by some external resource:

$$\begin{aligned}\dot{X}_1 &= X_1(2-0.8X_1-0.7X_2-0.5X_3)\\ \dot{X}_2 &= X_2(2.1-0.2X_1-0.9X_2-X_3)\\ \dot{X}_3 &= X_3(1.5-X_1-0.3X_2-0.2X_3).\end{aligned} \tag{3.1}$$

The non-trivial equilibrium is at 1, 1, 1 and it is locally stable; the eigenvalues are approximately -1.88 and $-0.009\,85 \pm 0.288i$. Yet for many initial conditions the system does not return to 1, 1, 1 but to 0, 0, 7.5, which is an equilibrium with only X_3 present. Figure 3.1 shows a small part of the space of X_1, X_2, X_3, the three-species equilibrium and a surface which divides initial densities that result in the species densities returning to 1, 1, 1 from those that lead to 0, 0, 7.5. I obtained this surface through simulation: the simulation was time-consuming, spanned only a small part of the possible initial conditions, and would have to be repeated if any of the parameters were changed. Is there an analytical method of deciding whether a system is globally stable?

3.2 GLOBAL STABILITY

A technique does exist to determine whether a system is globally stable. It involves finding a function known as a Lyapunov function. Demonstrating this global stability will illustrate why the existence of a Lyapunov function is often so difficult to determine for multispecies models and, consequently, why this approach has a very limited utility. The discussion will be facilitated by considering physical systems analogous to the biological ones. Tuljapurkar &

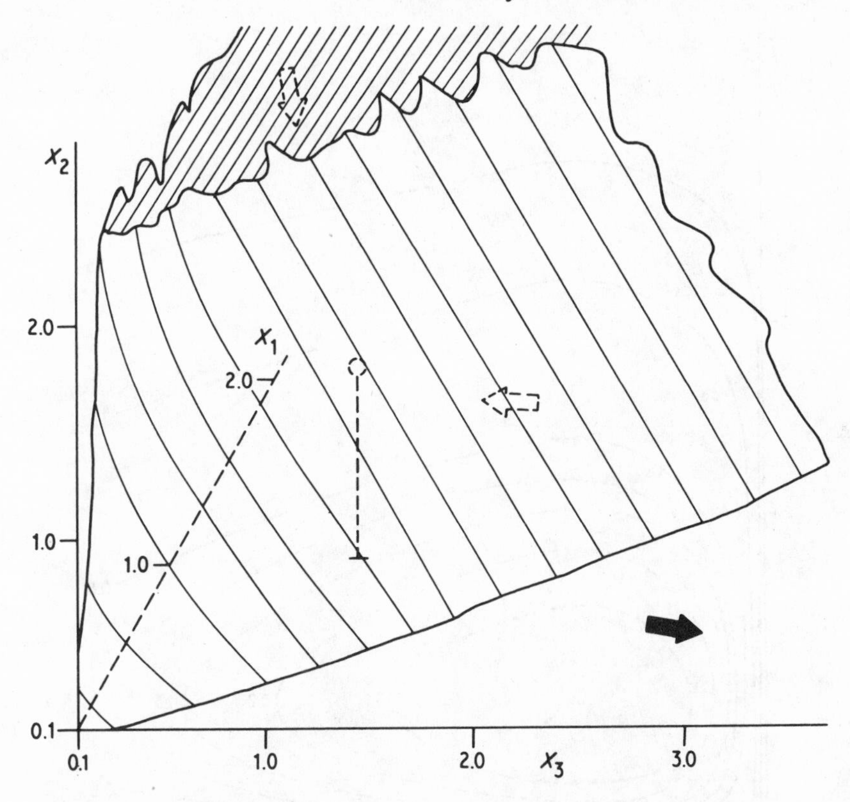

Figure 3.1 A locally, but not globally, stable system. The locally stable equilibrium involving all three species is at (1, 1, 1) and indicated by the 'ball on the stick'. Axes are the initial densities of the species. From inside the wedge-shaped object (broken at high values of X_2 – top of the figure – and at high values of X_1 and X_3 – top right of the figure), the densities return to the three-species equilibrium. From outside this volume the species densities move in the direction suggested by the solid arrow, to a one-species equilibrium at (0, 0, 7.5)

Semura (1979) have argued that such analogies are not always appropriate to population dynamics. However, the analogies are appropriate to the purpose of this discussion: to demonstrate why alternative definitions of stability under large perturbations must be examined.

Consider a predator–prey model:

$$\dot{X}_1 = X_1(2 - X_1 - X_2)$$
$$\dot{X}_2 = X_2(-0.3 + 0.3X_1). \tag{3.2}$$

The equilibrium is at (1, 1) and the eigenvalues are approximately $-0.5 \pm 0.477i$. The system is at least locally stable. Through simulation, I obtained the results shown in Fig. 3.2. The path of successive values of X_1 and X_2, called

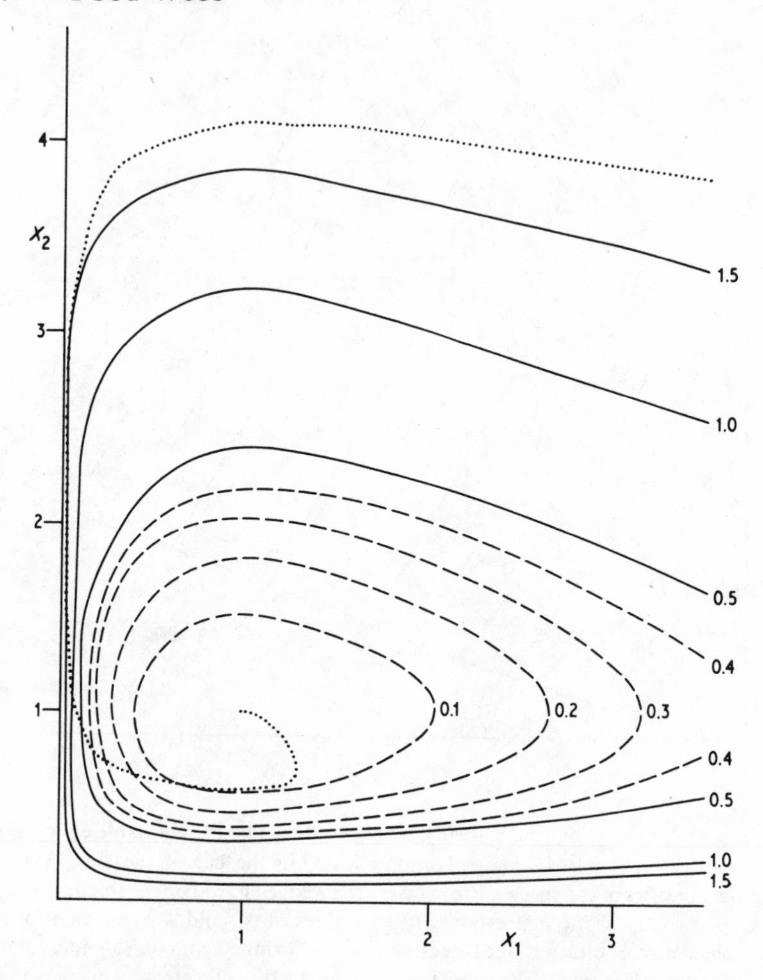

Figure 3.2 Densities of X_1 versus X_2 for a two-species predator–prey system discussed in the text. Equilibrium is at (1, 1). The dotted line is the path followed by the two species' densities from (4, 4) until it reaches equilibrium. Full and broken lines are contours, the former in intervals of 0.5, the latter of 0.1, derived from the Lyapunov function $g(X_1, X_2)$ discussed in the text. The critical feature of this figure is that the path of the two species densities always crosses the contours from higher values (the outside) to lower values (the inside).

the population trajectory, spirals inwards. This spiralling is expected because the eigenvalues have imaginary parts.

If the trajectories approach equilibrium from any set of initial densities, then the system is globally stable. Unfortunately, strict geometrical distance cannot be used to measure whether the system moves towards the equilibrium because there are times when the system moves geometrically away from it. The physical analogy is obvious: the domain of attraction is not a perfectly round 'basin'. But, if it were round, then concentric 'contours' could be drawn around the equilibrium. Because trajectories should always move 'downhill', it should be possible to derive conditions which showed that the trajectories always crossed from the outside of these circular contours to the inside. Even if the 'basin' is not 'round', it may be possible to find closed curves ('deformed circles') which the trajectories will always cross from outside to inside. Figure 3.2 shows some of these curves, which are given by

$$g(X_1, X_2) = 0.3[(X_1 - 1) - \ln(X_1)] + (X_2 - 1) - \ln(X_2). \tag{3.3}$$

These curves are the solution to $g(X_1, X_2) =$ a constant. (Their derivation will be discussed presently.) This constant is the 'height' above the equilibrium point.

Equation (3.3) has some important properties. First, it is not defined except for positive densities, since a logarithm (ln) cannot be taken of a negative number; negative densities of animals are not defined either. The equation is zero only when $X_1 = X_2 = 1$, that is, at equilibrium. Elsewhere the function is always positive and approaches infinity as either species density approaches zero or infinity. The conditions have obvious physical analogies. They imply that the bottom of the basin is at one, and only one, place, namely the equilibrium. As the species densities move further and further away from the equilibrium the sides of the basin become steeper, so steep in fact that the species densities can never leave the basin.

One other property is required for global stability: that the trajectories cross from the outside to the inside of these curves. In the analogy, the trajectory must always move downhill in the basin. Along any trajectory, the function g has the value $g(X_{1(t)}, X_{2(t)})$. The time derivative of g, $\dot{g}$, must always be negative if the trajectories are moving downhill. If g always decreases along any trajectory, yet is positive everywhere except at equilibrium, then the system is globally stable. To obtain the derivative of $g(X_{1(t)}, X_{2(t)})$ recall the result that, if $g = f(x(t))$, the function can be differentiated using the chain rule

$$\frac{\mathrm{d}g}{\mathrm{d}u} \cdot \frac{\mathrm{d}u}{\mathrm{d}t} = \frac{\mathrm{d}g}{\mathrm{d}t}. \tag{3.4}$$

Now Equation (3.3) must be differentiated and it is a function of both X_1 and X_2, where both are functions of time. Let $u_1 = X_1$ and $u_2 = X_2$; then $\mathrm{d}u_i/\mathrm{d}t = \dot{X}_i$. Now, replace the X_i in Equation (3.3) with the u_i and obtain $\mathrm{d}g/\mathrm{d}u_i$. Multiplying the terms to obtain $\mathrm{d}g/\mathrm{d}t$:

$$\dot{g}(X_1, X_2) = 0.3(1 - 1/X_1)\dot{X}_1 + (1 - 1/X_2)\dot{X}_2. \tag{3.5}$$

$\dot{X}_1$ and $\dot{X}_2$ are given by Equations (3.2) and can be substituted into (3.5). Simple algebra shows that several terms cancel, leaving

$$\dot{g}(X_1, X_2) = -0.3(X_1 - 1)^2. \tag{3.6}$$

Now, this is not always negative, for at $X_1 = 1$ it is zero. This means that the solution will not always move downhill; at $X_1 = 1$ it may 'remain' on a 'ledge'. Stability can still be proved in such cases using a theorem due to LaSalle (1960), the uses of which are discussed in Harrison (1980). The theorem states that if the equilibrium point X_i^* (for all i) is the only solution to $\dot{g} = 0$, then the system is globally stable. In terms of the analogy, the theorem excludes ledges where the solution might remain without returning to equilibrium.

Returning to the example, suppose that $\dot{g} = 0$. This means that from Equation (3.6) $X_1 = 1$ and, by differentiation, $\dot{X}_1 = 0$. Substituting this information into Equation (3.2) gives

$$0 = 1(2 - 1 - X_2), \qquad \text{that is, } X_2 = 1.$$

This is the equilibrium value for X_2. If a complicated function was obtained for X and if this was a possible trajectory for the system, then the function could be the equation describing the 'ledge' on which the trajectory could 'remain' without returning to equilibrium. In this case, there is no 'ledge' and system (3.2) is globally stable.

A formal statement of the conditions that ensure global stability is as follows:

In an n-species model let $\dot{X}_i = 0$, given that $\dot{X}_i = X_i f_i(X_1, \ldots, X_n)$ at $X_i = X_i^*$, where $X_i^* \neq 0$. The function $g(X_1, \ldots, X_n)$ is a Lyapunov function if it satisfies the following conditions:

(i) $g > 0$ for all X_i, except when all $X_i = X_i^*$.
(ii) For each i, $g \to \infty$ as $X_i \to 0$, or $X_i \to \infty$.
(iii) Along the trajectories of this system $\dot{g} \leqslant 0$.
(iv) $\dot{g} \neq 0$, except at $X_i = X_i^*$ for all i. This is relaxed by LaSalle's theorem, which states that except for the equilibrium trajectory the solution to $\dot{g} = 0$ must not be a trajectory.

The obvious question is: how was Equation (3.3) obtained? The choice of the Lyapunov function, g, cannot always be determined by a logical system of algebraic manipulations. And more than one function may exist for any system. Goh (1977) has discussed a Lyapunov function that fits all Lotka–Volterra models:

$$g(x) = \sum_{i=1}^{n} d_i[X_i - X_i^* - X_i \ln(X_i/X_i^*)] \qquad (3.7)$$

where the d_i are constants. *If* the d_i *exist* such that $\dot{g}$ is always negative except at X_i^* (when it is zero) then the system is globally stable. In the example $d_1 = 0.3$ and $d_2 = 1$, and $X_1^* = X_2^* = 1$; this gives Equation (3.3). The problem is how to obtain the d_i so that they satisfy condition (iii) above. For simple examples this is easy, but for more complicated examples it is not: there *is no certain way* of selecting the d_i. An intermediate result of Goh's leads to some theorems about the structure of the a_{ij} in the Lotka–Volterra models that ensure global stability. By and large these are not particularly useful for food web models, though May (1974b) has used the results to good effect in his studies on species packing. In addition, Harrison (1980) has shown that linear food chains (A eats B eats C, etc.) are globally stable, no matter how long they are.

In short, global stability is often hard or impossible to establish for biologically reasonable, but realistically complex food web models. However, this subject is currently undergoing active study. Interesting results may become available that provide insight into food web designs that are globally stable. Pending such results, I present in the next section another stability criterion which is analytically tractable yet involves large perturbations.

3.3 SPECIES DELETION STABILITY

Like the perturbations envisioned by global stability, species deletion is a relatively large perturbation. More critically, it is a persistent one. A system is said to be species deletion stable if, following the removal of a species from the system, all of the remaining species are retained at a new, locally stable equilibrium. Practically, species deletion stability can be calculated by deleting each species in turn from a model web and analysing the local stability of the resulting system.

Natural systems may lose species through emigration, epidemics, increased vulnerability to predators, catastrophic destruction by weather or other physical factors, the agencies of man, and many other factors. Species deletion, however, may model quite general, large perturbations to a system even though a species is not completely eliminated. Compare species deletion stability to global stability:

(i) Global stability implies a transient perturbation. Species deletion stability implies that a species is placed at zero density and stays there. In the real world this distinction becomes a matter of degree. When a species is perturbed but allowed to return, it may be effectively absent for long periods of time; so long, in fact, that other species may be lost from the system permanently or just long enough to give this impression. An example is provided by the four-species model with one predator (X_4), one herbivore (X_3) and two plants (X_1, X_2) shown in Fig. 3.3. The density of the herbivore depends on the ratio of two parameters, b_4 and a_{43}, in the predator's growth equation. This ratio is the same in the two runs illustrated by Fig. 3.3. All the other parameters and the initial conditions are the same, so that the species' densities return eventually to the same equilibrium densities. The parameters were selected so that X_3 has the most negative effect on the plant X_1 ($-a_{23} < -a_{13}$) but gains most benefit from the other plant species X_2 ($a_{32} > a_{31}$). In (a) the parameters b_4, a_{43} are ten times larger than in (b) and in system (a) the densities recover quickly. In (b) the recovery of the predator is slow, permitting the herbivore to keep X_1 depressed well below its normal equilibrium density

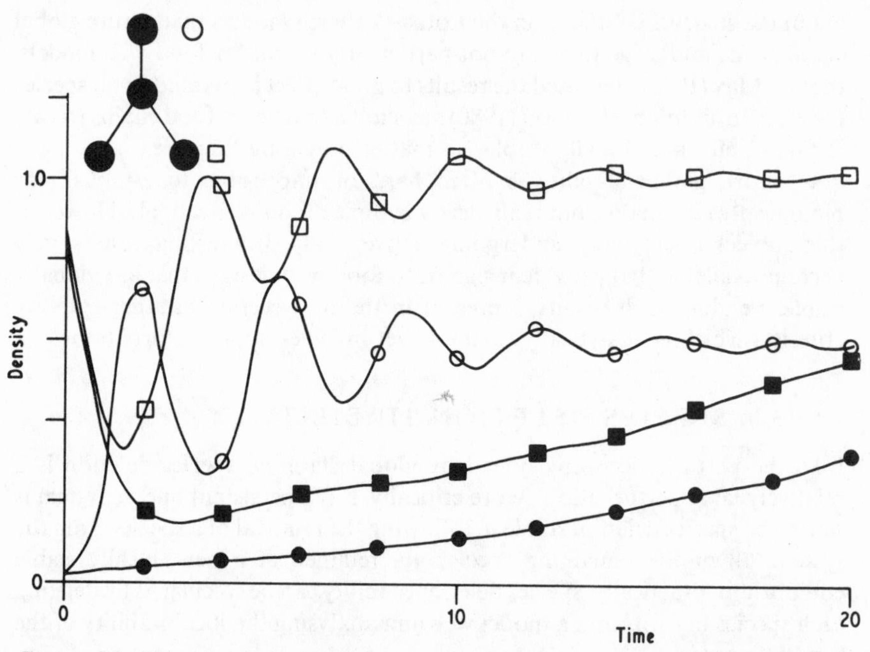

Figure 3.3 Two simulations of a four-species system (top left) with a carnivore, a herbivore, and two plant species. Two species densities are shown: the carnivore (circles) and one of the plants (squares). (a) In the first simulation (open symbols) the carnivore responds quickly to a perturbation. (b) In the second simulation (full symbols) the carnivore is slow to respond. Consequently, the herbivore (not shown) remains at high levels for long periods with a subsequent long depression of one of the plant species on which it feeds. (From Pimm, 1980c.)

for many generations. Under such circumstances X_1 might be lost from the community for any of the many reasons that make small populations susceptible to extinction. With even occasional perturbations to X_4, X_1 would remain rare and well below its potential equilibrium. This example is not contrived: it was suggested by Dayton's (1975b) study of an intertidal system where the top-predator, a starfish, was very slow to respond to changes in the density of its prey. In ecological time, the sluggish behaviour of system (b) suggests that large perturbations might best be modelled by the deletion of the predator, rather than by showing the system to be globally stable.

(ii) Global stability requires return from all population densities. But many populations have levels below which recovery will be difficult, or even impossible. No such threshold effects are included in Lotka–Volterra models. This may not be a serious source of error if densities are never small, but when large perturbations are considered, such inabilities to recover must be incorporated into the model. In ecological terms, a large perturbation may be exactly equivalent to removing a species completely. In such cases, species deletion stability may give a very good indication of whether a real system is globally stable.

I conclude that species deletion may be a reasonable approximation to the large perturbations which systems may sometimes suffer. It does not replace global stability as a criterion to be used in investigating food web structures, but it does provide an opportunity to examine the consequences of some large perturbations.

3.4 STABILITY IN STOCHASTIC ENVIRONMENTS

3.4.1 Two alternatives

In the previous discussions it was assumed that the perturbations are infrequent: the system is allowed to recover following a perturbation. This assumption is not altogether reasonable. The densities of natural populations are constantly buffeted by a variety of factors, including weather, and under these conditions the definition of stability is not simple. There are several criteria that can be applied and all are more complicated than I wish to discuss (see Turelli, 1978). Intuitively, however, it can be conjectured that populations which fluctuate the least are those most likely to persist. Large fluctuations will eventually bring densities to the low levels from which real populations will stand little chance of recovery. Even with this definition, ecologists seeking to model such stochastic effects face the dilemma of having two seemingly reasonable approaches which give diametrically opposite results. There are two arguments.

Alternative 1. The first argument is that species whose densities return quickly to equilibrium following a perturbation will vary less than those that return to equilibrium slowly (May, 1973b). Examples are given in Fig. 3.4(a). There are two simulations and both use a difference equation analogue (Equation (3.11) below) of a single-species model (Equation (2.17)). Each generation, a normally distributed random variable is added to the growth rate. The variance of these stochastic effects is the same in both simulations. The simulations differ in the rates at which the population returns to equilibrium. In one simulation (full symbols) there is a long return time, but in the other (open symbols) a short one. In the former case, particularly when the population level is small, it cannot easily recover. In the latter case, the population fluctuates less, recovers more quickly and does not reach the low population levels which, in the real world, would mean extinction for the species.

Alternative 2. The second argument is that the equilibrium density (carrying capacity) is, itself, variable. Species with high rates of return to equilibrium will track the fluctuations in equilibrium density. Such a species will have a more variable density than a species which responds more slowly

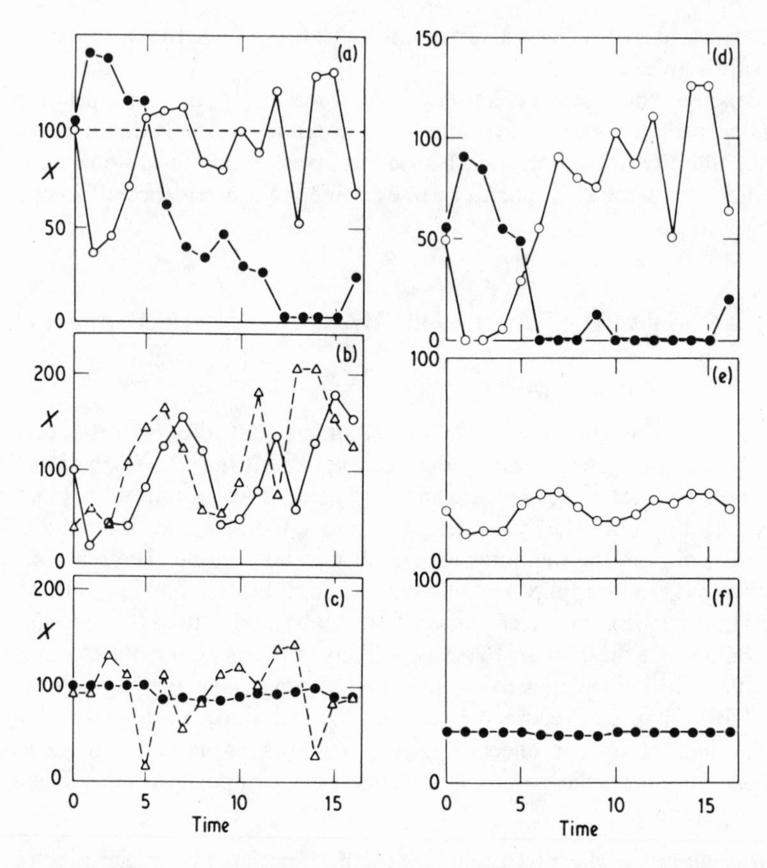

Figure 3.4 Simulations of a single-species population model. In (a) the carrying capacity (broken line) is constant; stochastic variation is added to the species' density. Full symbols: the rate of return to equilibrium, b, = 0.1; open symbols, b = 0.9. In (b) and (c) the carrying capacity (broken line connecting triangles) is variable. In (b) b = 0.9, in (c) b = 0.1. The simulations (d), (e), and (f) are the same as (a), (b), and (c) except that a predator has been added to the system. This increases the return times in each case. (From Pimm, 1982).

and 'waits' until more favourable conditions prevail (Turelli, 1978, Whittaker & Goodman, 1978, Luckinbill & Fenton, 1978, May, Beddington, Horwood & Shepherd, 1978). Examples are given in Figs 3.4(b) and (c) where the return times are exactly those in the simulations of Fig. 3.4(a). In these examples, the species with the higher rate of return to equilibrium (Fig. 3.4(b)) fluctuates more and, as a consequence, is more likely to become extinct.

Although the analysis of stochastic models is complex, it is possible to develop the preceding verbal argument analytically. A simple example using a one-species model

$$\dot{X} = Xb(1 - aX) \tag{3.8}$$

reasonably represents the argument presented intuitively above. Using Equation (2.19) to integrate Equation (3.8) over a unit time interval yields

$$X_{t+1} = (1/a)[X_t \exp(b)]/[X_t \exp(b) + (1/a) - X_t]. \tag{3.9}$$

Equilibrium at $1/a$ is approached at a rate dependent on b, $-b$ is the eigenvalue of this system and $1/b$ its return time. Under the first alternative, X_t is replaced with $X_t + z_t$, where z_t is a random variable. The population will have a smaller variance when b is very large (Equation (3.9) simplifies to $1/a$) than when $b = 0$ (Equation (3.9) simplifies to $X_t + z_t$). Under the second alternative, $1/a$ is replaced with $(1/a)(1 + z_t)$. When $b = 0$ the population will not fluctuate – it remains at X_t–but when b is infinite the equation simplifies to $(1/a)(1 + z_t)$.

These examples consider only a single trophic level and differ in their return times because the single-specific parameter (b) is different. However, it is easy to show that it is the return time of the system and not the parameter, b, *per se* that determines the populations' variability. In Figs 3.4(d–f) the same simulations presented in Figs 3.4(a–c) have been repeated exactly, except that another trophic level (a predator) has been added. The effect of adding another trophic level is to increase considerably the return times of both the systems with formerly short and long return times. (This result will be discussed in some detail in Chapter 6.) The effects are as predicted: in Fig. 3.4(d), where the random variation is added to the growth rates, fluctuations are much higher than before (Fig. 3.4(a)). In Figs 3.4(e) and (f), when random variation is added to the carrying capacity of the prey species, both simulations show only slight variation in numbers.

Regardless of how the changes in return times are brought about, long return times mean less variation in densities if random variation is added to the carrying capacity term but more variation in densities if random variation is added to the growth rates. Thus, the two alternatives make opposite predictions about how population variability varies with return times and, consequently, about the magnitudes of the eigenvalues to be expected in natural systems on which there are limits to how much variability can be tolerated without species extinctions. A comprehensive analysis of this topic is provided by Turelli (1978).

3.4.2 Evaluating the alternatives

Deciding *a priori* which alternative is likely to be the better description of a natural population is not easy. May *et al.* (1978) have suggested that certain patterns of variance in yield against harvesting effort for fish populations are in accord with alternative 1. There are experimental data to support alternative 2.

Luckinbill & Fenton (1978) experimentally manipulated the carrying capacity of two species of protozoa by varying the amounts of their food (bacteria) added to the culture. The species with the faster increase from below the carrying capacity was also the faster to decline from above the carrying capacity. As predicted by alternative 2, this species varied more than the other and was the first species to become extinct. Such results confirm that alternative 2 can describe the behaviour of populations, but it is still important to ask which alternative is the better description of the world outside the laboratory.

To compare the alternatives I used the British Trust for Ornithology's Common Bird Census (CBC) (Pimm, 1982), in which data on common farmland and woodland birds have been collected since 1962 (Battan & Marchant 1976, 1977, Marchant, 1978). The available data up to and including 1977 were analysed. The densities for each species are given on a relative scale set to 100 in 1966. The variation in densities was calculated as the coefficient of variation (CV). To estimate the rate of return to equilibrium (b) presents more difficulty. Clutch sizes are fairly well known (e.g. Witherby, Jourdain, Ticehurst & Tucker, 1938) but are much bigger than the realized rate. They also vary with habitat and population densities. I chose to estimate b from census data using the statistical model

$$Y_i = \beta_1 X_i + \beta_2 X_i^2 + \varepsilon_i, \tag{3.10}$$

where the Y_i are the densitites in year $n+1$, X_i the densities in year n, β_i the parameters to be estimated and the ε_i permit random deviations from the model. For models of this form it is a straightforward process to obtain the β_i given sets of variables for the X and Y (see, for example, Draper & Smith, 1966).

Consider the meaning of β_1. If Equation (3.9) is expanded as a Taylor series about equilibrium, it can be shown that $\ln(\beta_1 - 1)$ is an estimate of b. Now, populations reproducing only during a limited time each year may be modelled by a finite difference equation rather than a differential equation like (3.8), which has continuous birth and death processes. Reality may be somewhere in between the assumptions of the two models. Consider a finite difference analogue of (3.8), where $\dot{X}$ is replaced by $X_{t+1} - X_t$, that is

$$X_{t+1} = X_t + X_t(b - aX_t) = X_t(1+b) - aX_t^2. \tag{3.11}$$

It is apparent that β_1 estimates $1+b$ or, in other words, $\beta_1 - 1$ is an estimate of b. Equation (3.11) is only one possible analogue of Equation (3.8); Equations (3.9) and (2.15) are others. They are qualitatively similar but b, though a function of β_1 in each, is estimated in slightly different ways. It is the *sign* of the relationship between b and CV that is of interest, not the precise functional form. This suggests that β_1 provides a reasonable independent variable to be used in the analysis as some measure of the rate of return to equilibrium.

In both (3.9) and (3.11) $X_t = 0$ implies $X_{t+1} = 0$, and so there is no intercept (term in X^0) in the statistical model. The CBC data were collected in two habitats (farmland and woodland), but because farmland species occurred in woodland and vice versa the two sets of data cannot be considered independent. Two sets of data – farmland species plus additional woodland species and woodland species plus additional farmland species – are plotted in Fig. 3.5. The correlations between variability and the measure of the rate of return to equilibrium are negative. The data show that, as some function of the rate of population growth increases, so the variability of the population density decreases. This is the result predicted by alternative 1. Alternative 2 was rejected at levels of 0.02 and 0.07 for the analyses in the order listed above.

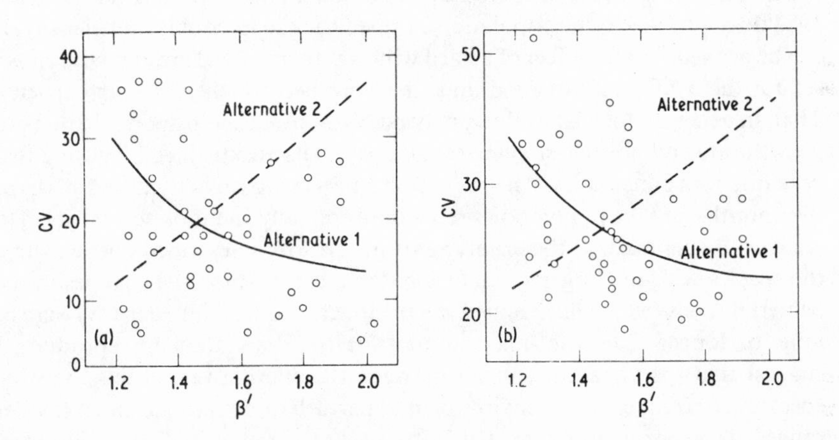

Figure 3.5 A measure of population variability (coefficient of variation, CV) versus a measure of the rate at which a population returns to equilibrium (β_1). (a) Farmland birds plus additional woodland species and (b) woodland birds plus additional farmland species. Two model alternatives discussed in the text are fitted to these data; in both (a) and (b) alternative 1 is adequate to describe the data, but alternative 2 is not. (From Pimm, 1982).

Dr Michael Turelli has kindly supplied an analysis of this problem obtained by linearizing the stochastic models about the equilibrium density (details are in Pimm, 1982). The analyses relate the species' CV to its growth rate, b, and the underlying environmental variation, s^2, for both alternatives. This permits an estimate of the average s^2 to be obtained under both alternatives. In turn, this permits the expected functional relationships between CV and β_1 to be calculated. These relationships are also shown in Fig. 3.5 where it is clear that, though the relationship derived for alternative 1 is an adequate description of the data, that derived for alternative 2 is not.

A possible criticism of these results stems from the potential interdependence of the dependent (CV) and independent (β_1) variables when both are estimated from the same data. Moreover, the assumptions for the regression model might be violated when serial data are used. To investigate the behaviour of the parameter estimates obtained from the same data, I performed computer simulations of alternatives 1 and 2. Certainly, the estimates of the parameters were found to be biased (Pimm, 1982), but the biases did not alter the conclusion that, at least for birds, stochastic effects were modelled better by perturbing growth rates than carrying capacities. Indeed, these analyses suggested that the significance levels (0.02, 0.07) are conservative: alternative 2 can be rejected with greater confidence than they suggest.

Why should this be so? An example may be informative. The density of the Song Thrush in farmland (Fig. 1.2(a)) fell drastically during the unusually cold winter of 1962–3. Most of the variation in its numbers comes from the years 1963 through 1967 as the population climbed back to its equilibrium density. It can be argued that the effect of a hard winter is to cover the ground with snow, reduce the availability of food and therefore perturb the carrying capacity. That, in general, stochastic effects may be better modelled by perturbations to growth rate and not the carrying capacity is perhaps explained by noting two very different time scales. The reduction in carrying capacity lasted perhaps two months, but the population breeds only annually and took several years to recover. By contrast, in the experiments on protozoa described above, where the species with the larger value of b became extinct first, the changes in the bacterial levels (which determined the protozoa's equilibrium density) were as long, or longer than, the time required for the population to reproduce. I suggest that the apparent conflict between the alternatives reflects the wide spectrum of frequencies of environmental perturbations and the much smaller range of response times available to natural populations in a seasonal environment. Population models allow continuous population responses and continuous perturbations, and this is unrealistic. Populations will recover from large but transient perturbations to equilibrium, much like the Song Thrush, with the most rapidly changing populations recovering the fastest and thus varying the least. More persistent changes to equilibrium may well lead to species which respond more slowly having lower variation. In short, the results indicate that perturbations to at least bird populations may be modelled better by changes to growth rates than to carrying capacity. The possibility that some populations are subjected to low-frequency perturbations (i.e. ones lasting several generations), where the opposite would be true, is not excluded.

3.5 OTHER STABILITY CRITERIA

So far I have been concerned with whether or not species densities return to equilibrium following a perturbation. Such a dichotomy is too simple, as population models permit other, more complex behaviours. A readable review of this topic can be found in May (1979) and I shall discuss these complexities only briefly.

An equilibrium may be unstable, but the populations may persist indefinitely by cycling about that equilibrium. Consider, for example, the finite difference Equation (3.11) for a single species. The equilibrium occurs at b/a; linearizing about this point gives the single eigenvalue of $1-b$. The parameter, b, the growth rate, must be larger than zero to have any biological sense, but there is nothing to prevent it from being greater than two. In such a case, the eigenvalue is less than minus one and the equilibirum is unstable. Simulation of such a system shows that the population does not inevitably become extinct for values of b greater than two. Rather, the population oscillates between values above and below the equilibrium. These oscillations become increasingly erratic as b increases and eventually become 'chaotic', which, as May (1974a) defines it, is seemingly random behaviour. May (1979) gives an interesting example of a laboratory population of rotifers whose numbers become increasingly erratic as temperature (and thus probably b) increases.

Complicated cycles are even easier to generate with two or more species and they are not restricted to difference equations. In two-species systems, *limit cycles* can be easily obtained with equations only slightly more complicated than those discussed so far. Limit cycles have the property that, although the equilibrium is unstable, the trajectories from all starting points move towards a fixed cyclical path. It is this cycle of population densities, rather than a fixed equilibrium, that attracts the trajectories. Populations which cycle repeatedly, like the Canadian Lynx (Fig. 1.1(b)), are probably best described by equations with this property.

A definition of stability that encompasses both stable equilibria and attractor trajectories is given by Hallam and co-workers (Hallam, Svoboda & Gard, 1979, Gard & Hallam, 1979, Hallam, 1980). Hallam defines as *persistent* any system that does not include a trajectory which leads to one or more species becoming extinct. Persistent systems can show a variety of behaviours and include systems that have more than one possible stable equilibrium (see May, 1979).

Another perspective of stability is that of Goh, Vincent and Anderson (Vincent & Anderson, 1979, Goh 1975, 1976). They have taken the view that some of the assumptions in the stochastic studies are unwarranted. In particular, these studies require the frequencies of the perturbations to be uniformly distributed. Such perturbations are called 'white' noise (i.e. they are without a dominant frequency or 'colour'). Yet perturbations in the real world have dominant frequencies, for example, winters occur once a year. Consequently, real perturbations may be considerably more harmful to a model system than white noise. Vincent & Anderson (1979) define the *reachable set* as all those combinations of species densities that can be reached over any length of time, even if the perturbations act in their most malevolent way. When the reachable set does not include a zero species density, then no species can be lost from the system. Such systems are deemed *invulnerable* to perturbations of a particular size. Goh (1976) shows that vulnerability generally decreases with decreasing return times, though this result is not inevitable (Vincent & Anderson, 1979).

3.6 SUMMARY: MODELS AND THEIR STABILITIES – IS THERE A BEST BUY?

The objective in developing the material in the last two chapters was to provide the machinery to decide which food web structures are probable and which are not. I argue that probable structures in the real world are those that have features which lead to dynamical stability in models. An immediate problem should be obvious. I have presented three model formats (differential equations: Lotka–Volterra and donor-controlled; difference equations), four definitions of stability (local, global, species deletion and stochastic) and touched on several other possibilities. Moreover, within the stochastic definition of stability there were two alternatives with diametrically opposite predictions about the magnitude, though not the sign, of the largest eigenvalues to be expected from real-world systems.

Is it necessary to investigate all the possible combinations of model formats and definitions to stability to determine their separate consequences to food web structure? Notwithstanding the sheer size of this task, some of the definitions of stability are difficult or as yet impossible to apply to models as complex as those of food webs. There are several possible strategies. Investigations of some combinations of model formats and stabilities are inevitable. Some of these combinations give essentially similar predictions. Yet others give different predictions about possible food web shapes; for these it must be decided whether certain of the combinations are biologically less appropriate than others. Finally, it is not possible to derive any biological predictions from some stability criteria. This does not mean they are biologically inappropriate; in such cases I seek a similar criterion that is easier to work with, at least as a temporary measure.

Is there a 'best buy'? Is there a biologically most reasonable combination of model format and stability definition from which to predict expected food web structures in the real world?

3.6.1 Definitions of stability

The definitions of stability can be easily classified by the size and the frequency of the perturbation that each definition envisions:

	Magnitude of perturbation	
Frequency	*Small*	*Large*
Infrequent	local	global species deletion
Frequent	stochastic	(mathematically intractable)

It is usually impossible to determine the global stability of a complicated food web model. But I have argued that species deletion may have many of the features of the large perturbations that the real world may sometimes suffer. Stochastic definitions are only tractable in multispecies models when the system is linearized. This implies that they are only certainly valid for small perturbations. I know of no technique combining both stochastic and large perturbations that is applicable to multispecies models.

Can a decision be made on whether small or large perturbations are more important in shaping the structures of natural food webs? There is likely to be a range of perturbations that destroys some systems but leaves others intact. Perturbations that are too small may have no effect, while those that are too large may destroy all systems. All natural systems suffer and survive small perturbations: models should be at least locally stable. In Chapter 5, data are given to demonstrate that natural systems are rarely species deletion stable. If natural systems do not withstand large perturbations, it is unreasonable to expect real food webs to have *only* those features that, in models, ensure resistance to large perturbations. After Chapter 5, I shall concentrate on small perturbations. Here, the strategy is to use model predictions to reduce the possible number of combinations that must be investigated. The emphasis on local stability is not entirely satisfactory because I have not answered Goh's (1977) comment exemplified by Fig. 3.1: local stability analyses do not determine whether the domain of attraction is merely small (which is acceptable) or infinitesimally small (which is not). The ideal would be a knowledge of the size of the domains of attraction of real and model systems, since these should be neither infinitesimal nor global. Such an ideal would be both theoretically and experimentally difficult to obtain.

Can a decision be made on whether frequent or infrequent perturbations are more important in shaping the structures of natural food webs? For both, stability requires the eigenvalues of the system to be negative. But for frequent perturbations there are further constraints on the magnitude of the eigenvalues. These constraints are in opposite directions depending on which of the two alternative models of stochastic effects is used. I have been able to show that, at least for birds, alternative 1 is better than its rival: stability is enhanced by strongly negative eigenvalues. From this I suggest the tentative conclusion that stability in stochastic environments is most likely when the largest eigenvalue is as negative as possible – 'negative' to ensure stability in even a deterministic world; 'as possible' to counter the destabilizing effects of frequent perturbations. In short, both the sign and the magnitude of the largest eigenvalue of a model system should be considered. This I shall do. But it should be clear that stochastic and deterministic alternatives are merely quantitative gradations of the same problem.

To summarize, the 'best buy' is to determine the largest eigenvalue of a system linearized about equilibrium. This information is appropriate to either frequent or infrequent small perturbations. I shall continue to consider large perturbations, however, until data have been presented to show that they usually destroy natural systems and are therefore unlikely to shape food web structures.

3.6.2 Models

Is there a best buy in models? Three formats were presented in the last chapter. Donor-controlled and Lotka–Volterra dynamics will be seen to have very different consequences to food web structures. These two formats also make different predictions about the effects of removing predators from systems. Using these predictions, I shall show in Chapter 5 that donor-controlled dynamics are of relatively limited application, though there are some special interactions modelled by them. I shall place more emphasis on Lotka–Volterra models but will indicate when their predictions differ from donor-controlled ones. In contrast, finite difference and Lotka–Volterra models place similar restrictions on possible food web shapes.

Are these three model formats enough to capture sufficient of the dynamics of real populations to be meaningful? And, if more complicated dynamics are required, are more comprehensive definitions of stability needed? There is a widely held belief among ecologists that Lotka–Volterra models are so obviously wrong that any prediction made from them is inevitably incorrect. I do not agree. Whether Lotka–Volterra models are 'obviously wrong' or not is a matter of perspective. In introducing Lotka–Volterra models I drew attention to two perspectives. The first was that the assumptions behind the models could be justified biologically. If ecologists view the equations in this way and as equations describing the global dynamics valid for all the achievable population densities, then there is no doubt that the equations are inadequate. But there is another perspective. It is that the parameters for the models represent terms in the Taylor expansion of the actual and potentially more complicated functions that describe the populations' growth. In this case, near equilibrium the equations are likely to be good descriptions because they are general approximations to many different complicated models, and not because the assumptions for the models are particularly accurate over large ranges in population densities. Were large perturbations to be common then the equations would be invalid. Much more care would be needed in formulating the models so that they would accurately describe the dynamics over wide ranges of population densities. I have already anticipated the result that small perturbations seem most effective in shaping the structure of natural food webs. If most populations remain near equilibrium, then the models discussed in this chapter are likely to be perfectly adequate for an investigation of food web structures.

Of course, even for detailed descriptions of the dynamics of one or two species, not all additions to model sophistication change the dynamics by much – as Hassell (1979) has shown in his analyses of insect host–parasitoid systems. Nor, as Bellows (1981) has shown in his 'consumer's guide' to single-species models, are more complicated models inevitably better descriptions of the real world than simple ones.

There is one remaining problem. I have argued that the structures that lead to instability will not be features of the real world. Yet systems with unstable equilibria may show cyclical changes in population densities with all species persisting. In the real world, cycles of large magnitude are not common; those of the Lynx (Fig. 1.2(b)) in the forests are conspicuous, but rare exceptions. But suppose cycles of small magnitude were common. I cannot exclude these by such arguments. No population density is absolutely constant and cyclical changes of small amplitude would be almost impossible to detect. In all the population models with which I am familiar and where conditions for stability and cyclical behaviour have been determined, the set of parameters that lead to small-amplitude cycles form a 'halo' around those that lead to local stability. Features which make local stability more probable are usually those which make both local stability and small-amplitude cycles more probable, and vice versa. Using the criterion of local stability for these systems will certainly lead to incorrect *quantitative* predictions about which food web structures are likely. But *qualitative* predictions (e.g. whether this structure is more probable than that one) are likely to be in the right direction.

4 Food web complexity I: theoretical results

4.1 INTRODUCTION

For centuries, biologists have been leaving their north-temperate, cold and species-poor homes to marvel at the extraordinary richness of species in the tropics. Nor has the contrast between the often capricious temperate climates and the warm and seemingly constant tropical ones escaped their notice. Not surprisingly, some of the earliest attempts to relate food web attributes to ecosystem function are embodied in the idea of some relationship between stability and complexity. But species richness as a measure of complexity and a benign, aseasonal climate as a measure of stability is only one of the many possible pairs of definitions. The hypothesis that complexity is related to stability causes contention, in part, because the definitions of stability and complexity are sufficiently vague to permit a plethora of intepretations. Elton (1958, pp. 145–153) gives six lines of evidence for his assertion that more complex ecosystems are more stable. His examples seem highly heterogeneous in the light of twenty years of hindsight. They contain, for example, comments on the cyclical behaviour of single-species population models,a statement that tropical systems are free from pest outbreaks and a discussion of how species-poor island faunas are susceptible to man's attempts at introducing new species. But most contention stems from recent theoretical studies because these lead to the opposite hypothesis, namely, that complex systems will be less stable than simple ones.

The 'complexity begets stability' idea was treated by MacArthur (1955) and Elton (1958). MacArthur's hypothesis was a simple and appealing one: the more choice energy has in flowing through a food web (complexity), the less will be the change in species densities (stability) when one or more species have an abnormally high or low abundance (a perturbation).

Some of the growth of the problem since the mid-1950s can be seen from the reviews of May (1973b; largely theoretical aspects), Goodman (1975; the idea that tropical systems are more stable than temperate ones) and several papers in the Proceedings of the 1st International Congress of Ecology (May, 1975, Orians, 1975, Margalef, 1975), plus a reply to these papers (McNaughton, 1977). All deal with possible meanings of stability and complexity. My interest in this problem is much more limited: I emphasize its aspects as they relate to food web design.

Consider two food webs of intertidal communities, with the first having about one-half the number of species of the second, as shown in Fig. 4.1. Interestingly, the former is from further north, 49 °N, than the latter, 31 °N. Both are from Paine (1966). Which system is the more stable? Observe that both are stable in a sense: their species have persisted for considerable periods of time. Two possibilities arise from this observation.

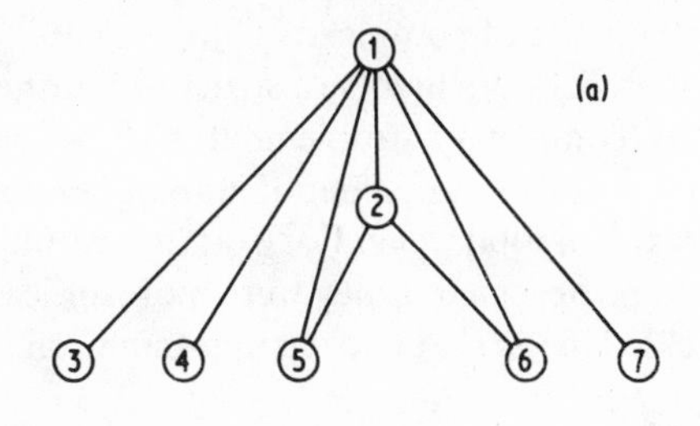

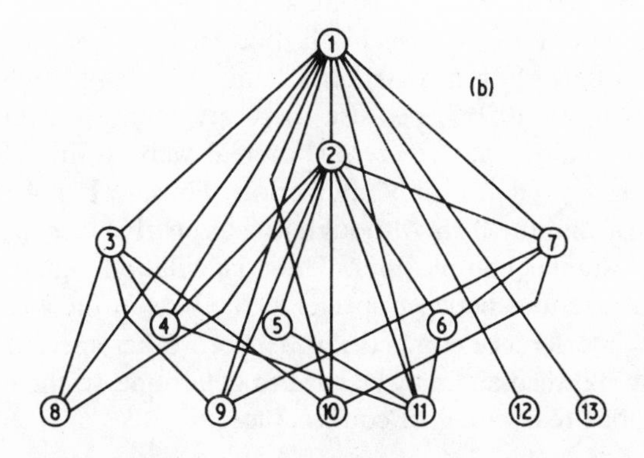

Figure 4.1 Two intertidal food webs: (a) Mukkaw Bay, Washington at 49 °N and (b) the northern Gulf of California at 31 °N. (a): (1) *Pisaster*, (2) *Thais*, (3) Chitons, two species, (4) Limpets, two species, (5) a bivalve, (6) acorn barnacles, three species, and (7) *Mitella*; (b) : (1) *Heliaster*, (2) *Muricanthus*, (3) *Hexaplex*, (4) *Morula*, (5) *Cantharus*, (6) *Acanthina tuberculata*, (7) *A. angelica*, (8) Columbellidae, five species, (9) bivalves, 13 species, (10) herbivorous gastropods, 14 species, (11) barnacles, three species, (12) Chitons, two species, (13) a species of brachipod. After Paine (1966).

(i) For the first possibility it is necessary to ask: is one system *relatively* more stable than the other? This question demands a definition of relative stability and such definitions are scarce in both theoretical and experimental studies. Return time is one measure of relative stability. Systems with short return times can be considered more *resilient* than those with long return times. In

Chapter 10, I will show that although more complex webs are less likely to be stable than simple ones, stable systems are more resilient, the more complex they are. This shows that a relative measure of stability – resilience – does not vary with complexity in the same way as a qualitative measure, namely, whether or not the system returns to equilibrium. Let me postpone this discussion because resilience is a feature of how an ecosystem functions (in the face of a perturbation) and the relationship between structure and function is much broader than the issue of complexity.

(ii) The second possibility does involve a structural (rather than functional) aspect of the stability–complexity problem. Real food webs must contain stable populations. Does this requirement of stability impose upper or lower bounds on the complexity of food webs? For example, would webs like those in Fig. 4.1 be expected to be observed if they were more, or less complex? This is the question to be tackled in the next two chapters, and I start with a definition of complexity.

A common measure of complexity is *connectance*, that is, the actual, divided by the possible number of interspecific interactions. For an example, consider a three-species food chain, with X_3 eating X_2 eating X_1. There are three pairs of interspecies interactions of which all but two (a_{13}, a_{31}) are zero. The connectance is 0.67. For a six-species model of two similar chains, the connectance is lower (0.267) because there are many possible interactions (those between chains) that are zero. For real webs, a line connecting two species indicates a predator–prey interaction. Thus, in Fig. 4.1, the connectances are (a) 0.38 and (b) 0.36. Although (b) has nearly twice the species of (a), the webs have similar complexities. These calculations ignore the possible competitive interactions between species at the base of the web for resources and the direct interference competition between other species. Such interactions are rarely specified and may be hard to determine, so there is often some uncertainty about real values of connectance.

4.2 BOUNDS ON FOOD WEB COMPLEXITY: LOCAL STABILITY

4.2.1 Preliminary results

The methods used to investigate the relationship between stability and complexity are simple in principle. A web of a given size and complexity is chosen and represented by the equations discussed in Chapter 2. Then the decision on whether the system is stable is made by using the machinery also described in Chapter 2. Indeed, all that must be done is to specify the elements of the Jacobian matrix and then find the largest eigenvalue of this matrix. This process can then be repeated to find the proportion of models, of a given complexity, that are stable.

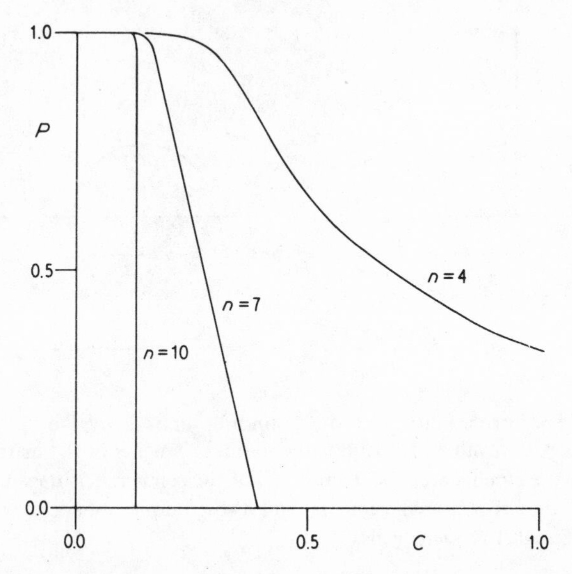

Figure 4.2 The probability (P) that a model of a given connectance (C) will be stable for model webs with different numbers of species (After Gardner & Ashby, 1970).

Figure 4.2 shows the results obtained by Gardner & Ashby (1970). As connectance (C) increases, the proportion of stable models decreases slowly to a critical value whence it drops rapidly. The critical value of connectance becomes smaller as more species are added to the web. May (1972) was able to derive this result analytically, using a result on the distribution of eigenvalues. Suppose that the elements of a matrix, $\boldsymbol{D}$, are symmetrical ($d_{ij} = d_{ji}$) and are drawn randomly from a statistical distribution with a mean of zero and a standard deviation of s. This parameter, s, is called the interaction strength and is a measure of the magnitude of the d_{ij}. These can be either positive or negative, so s measures how far these numbers are from zero by taking the mean of the square of the d_{ij} (which is s^2); the squared terms are positive, irrespective of the actual signs of the d_{ij}. When the number of elements is large, the distribution of the eigenvalues can be considered a continuum. The frequency distribution of the eigenvalues is a semicircle with radius $s(n)^{1/2}$, centered on zero; n is the number of species (Wigner, 1959, Mehta, 1967). Subtracting a constant (k) from all the diagonal entries of $\boldsymbol{D}$ will shift the semicircle to the left by an amount equal to that constant (Fig. 4.3).

From the figure it is obvious that if $s(n)^{1/2}$ is larger than k there is a finite chance of one of the n eigenvalues being positive (and hence the system being unstable). And, for large n, there will be a high probability of obtaining at least one positive eigenvalue. If unity is subtracted from each diagonal element of the matrix, the system will be stable if

$$s(n)^{1/2} < 1 \tag{4.1a}$$

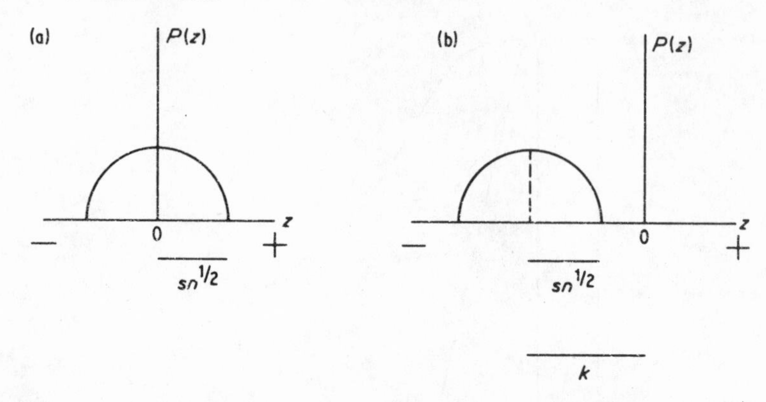

Figure 4.3 The probability ($P(z)$) of finding an eigenvalue of a real, symmetric matrix with a value of z. In (a) all diagonal elements of the matrix are zero, in (b) all diagonal elements are $-k$. The radius of the semicircle is dependent on s (the magnitude of the off-diagonal elements) and n (the number of species in the system). (After Post, Shugart & DeAngelis, 1978.)

and usually unstable if

$$s(n)^{1/2} > 1. \tag{4.1b}$$

Ecologically interesting matrices are unlikely to be symmetrical, nor does each species in a web interact with every other species. May (1972) allowed a proportion of the matrices' elements $(1-C)$ to be zero. McMurtrie (1975) performed computer investigations with asymmetrical matrices and also chose a proportion of the elements to be zero. The diagonal elements were fixed at -1. Both authors obtained the following conditions: the system will usually be stable if

$$s(nC)^{1/2} < 1, \tag{4.2a}$$

and usually unstable if

$$s(nC)^{1/2} > 1. \tag{4.2b}$$

To summarize, increased model complexity leads to a sample of randomly constructed webs containing fewer stable systems. If complexity is constant, increased species numbers will also lead to fewer stable models. The ecological implication of this result is that webs should be relatively simple: complex webs will not likely persist since some of their constituent species will be lost.

This result contradicted expectations, but it has numerous assumptions that are biologically untenable. First, consider how the models are formed. Selecting entries randomly over prescribed intervals may invoke an image of randomly throwing species together in some melting pot. What emerge are only those systems that are stable. This denies species the chances to evolve and so 'fine-tune' their parameters to maximize the chances of their system persisting. Saunders (1978) has objected to this assumption and suggested that 'fine-tuning' may permit even the most complex model to be stable. Incidentally, it is not obvious that evolution stabilizes species interactions. Rosenzweig (1973) and Rosenzweig & Schaffer (1978) have shown that just the opposite is true in some cases.

The interpretation of the model-building process I favour is that developed by May (1975). In the real world, model parameters are not constant, for they depend on a variety of influences external to the system. How efficiently a predator captures its prey, for example, may depend on the exact density of cover in which the prey hides, how active the prey is, and so on. Parameters are probably best considered as being distributed over a range of values and there is a limit to how finely they can be tuned. When only a small subset of the possible combinations of parameters is consistent with stability, the actual parameters may remain outside the subset for much of the time and the system will be unstable (Fig. 4.4(b)). When the proportion of possible parameter values that leads to stability is large, the actual parameter values may remain within this subset for most of the time giving a stable system (Fig. 4.4(a)). Parameters vary over space, as well as time, and the preceding arguments can refer to the chances of finding a particular system over space also.

In short, when the subset of parameters consistent with stability is small, the system can be considered to be dynamically fragile (May, 1975) and unlikely to be common in the real world.

This might suggest qualitative stability as a useful criterion to apply to food web models. Recall that for qualitative stability, all parameter values of a given sign structure ensure stability. Thus, qualitatively stable structures are those most likely to persist in the real world. But these structures are not the only ones to be expected. Some common interactions violate qualitative stability.

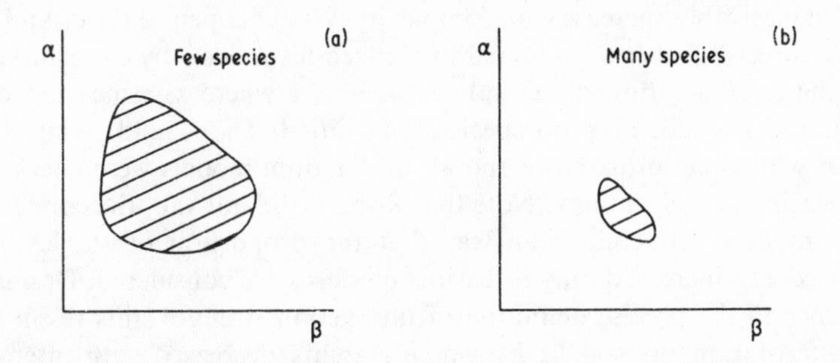

Figure 4.4 A hypothetical and two-dimensional example to illustrate 'dynamical fragility'. In general only certain combinations of parameter values will result in a model's stability. For the two parameters α, β this is indicated by the shaded area. Parameters in the real world will not be fixed but will vary over limits implied by the axes of the figures. On average system (a) is more likely to be encountered than system (b) because the parameter values consistent with stability are a greater proportion of those possible in (a) than in (b). (After May, 1975.)

Although the subset of parameters which yield stability may be relatively large, it may not be infinite. For example, direct competitive interactions violate qualitative stability yet still permit large ranges of parameters consistent with stability. In short, web features that give qualitative stability should occur in the real world but need not necessarily be predominant.

There are still three potential problems with the result embodied by Equations (4.2a, b): (i) The models may not involve positive population densities for all of the species. When all equilibrium densities are positive, the model is called *feasible* (Roberts, 1974). (ii) Real food webs do not have random interactions either in terms of how large the parameters are or in terms of which species interact. (iii) There are unrealistic assumptions about the sign and magnitude of the terms in the models. In particular, the models assume that all species populations are regulated, at least in part, by processes independent of other species and that predators have a strong, controlling effect on their prey.

4.2.2 Feasibility

Feasibility is an obvious requirement. Unfortunately, the studies discussed earlier did not evaluate it. Rather than specifying the individual terms of some particular model (e.g. the b_i, a_{ij} in the Lotka–Volterra models), the studies specified the elements of the Jacobian matrix (the product of the a_{ij} and the equilibrium density of the X_i^*). This implicitly fixes the X_i^* to be positive, because for Lotka–Volterra models some b_i can be found that, given the particular values of a_{ij}, will lead to a positive value. But what if constraints on the b_i are much more difficult to satisfy than those on stable systems? Then the earlier result might be reversed: within the set of feasible models, stability could, conceivably, increase with complexity. What happens if the a_{ij} and b_i are chosen direcly? Roberts (1974) parameterized his models by choosing the $b_i = 1$, the $a_{ii} = -1$ (for all i), and the $a_{ij} = \pm z$ where z, a measure of the interaction strength between species, was varied. The + and − signs were chosen with equal probability and all the random models were checked for both feasibility and stability. Note that Roberts did not vary the connectance of his models. His conclusion was that the proportion of stable models increased as z increased only if feasible models were considered. Despite the difference in the precise definition of interaction strength, this result is the opposite of Equations (4.2a, b), where stability decreased with interaction strength.

Gilpin (1975a) and Goh & Jennings (1977) repeated Roberts' computer experiments and showed that his result critically depends on the method of selection of the parameters. Gilpin chose $b_i = \pm 1$ and $a_{ii} = -1$ but kept Roberts' assumption of $a_{ij} = \pm z$ with the + and − signs again occurring with equal probability. With these assumptions, the stability of the models decreased with increasing z, suggesting that Roberts' results are fragile at best.

A detailed discussion of why Roberts' results are different is given by Post, Shugart & DeAngelis (1978). Here, it is sufficient to echo Gilpin's (1975a) comments. Roberts effectively assumed each of his species to be autotrophic, that is, each species could grow in the absence of others ($b_i > 0$). There is neither predation nor competition for resources in this model, only interference and facilitation, each equally likely, between species on a trophic level. Mutually positive or negative interactions are reasonable but the other possibilities are not. Gilpin asks:

> 'Why should the members of one species facilitate the members of a species that are interfering with them?'

In short, Roberts' results are a special case and not a very plausible one and are uncertain evidence for stability increasing with complexity.

4.2.3 Random interactions

The interactions in real food webs are not random in either sign or magnitude. The signs are particularly crucial. I know of no cases, in the real world, with loops of the kind 'species A eats species B eats species C eats species A' and excluding these loops prevents certain sign combinations. Lawlor (1978) has calculated the probability that a randomly organized community, with n species, will not have any three-species loops among any combination of three species. For $n = 10, 20, 30, 40$ and 60, the probabilities are approximately 10^{-2}, 10^{-16}, 10^{-56}, 10^{-137}, and 10^{-472}. Most randomly assembled communities are ecologically unreasonable: they have loops. Moreover, this is only one of a large number of possible sign restrictions that would ensure the model webs were 'reasonable'. Lawlor continued by suggesting an important possibility. He kept the product of connectance and the number of species constant (and equal to ten). If interaction strength was about 0.31, then the majority of Lawlor's systems would be expected to be stable, using the criterion of Equation 4.2. The frequency of systems without three-species loops was then recalculated. The probabilities of finding reasonable ecological systems were much larger (for $n = 10, 20, 30, 40$ and 60 they are 0.02, 0.573, 0.840, 0.927 and 0.977). More critically, the probability of getting a reasonable system increases with the number of species in the system. This shows that stable and reasonable models must lie together in the multidimensional space of possible parameters (that is, if a set of parameters ensures 'reasonableness' it will likely ensure stability and *vice versa*: Fig. 4.5(a)).

Lawlor's suggestion was this: as the number of species increases it is possible that the proportion of reasonable models that are stable might also increase. I shall show in the remainder of this section, for an admittedly small and restricted set of reasonable models, that this is not so (Fig. 4.5(b)). Reasonable models, however, are much more likely to be stable than random ones for any given combination of species, interaction strengths and connectance.

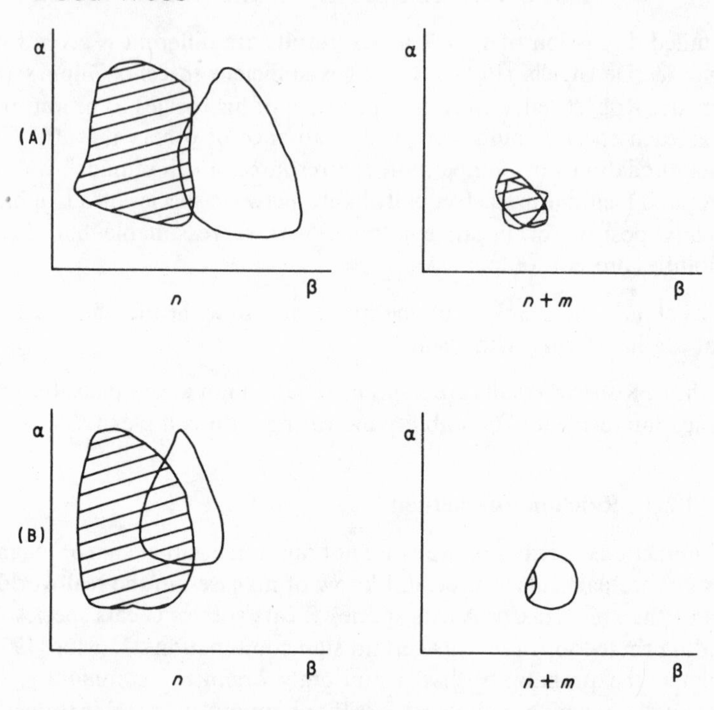

Figure 4.5 A two-dimensional representation of a multi-dimensional possibility suggested by Lawlor (1978). The sets of parameters that ensure stability (shaded area) and that are biologically reasonable (open areas) both decrease as the number of species in the system increases (from n to $n + m$). Both (A) and (B) are consistent with Lawlor's result that, among stable models, the proportion that are reasonable increases with increasing numbers of species. (A) It is possible that as the number of species increases the proportion of reasonable models that are stable also increases. Results of Pimm (1979b) suggest that possibility (B) is more likely, namely, that the proportion of reasonable models that are stable actually decreases with increasing number of species.

I chose six and eight species models with a restricted, yet biologically reasonable structure (Pimm, 1979b). These models excluded loops, predators without prey and singular systems. In both sets the species were organized into two chains (of three and four species respectively) as shown in Fig. 4.6. Each chain can be viewed as a plant, herbivore, carnivore and top-carnivore. Only the plants had densities that were limited by factors external to the system ($a_{ii} < 0$). (This assumption will be discussed in the next section.) I increased complexity by adding restricted predator–prey interactions, for instance, a predator could only prey on a species that was at a lower trophic level. The elements of the Jacobian matrix were chosen directly thus avoiding the problems of assessing feasibility.

These models are only a tiny subset of all possible six- and eight-species

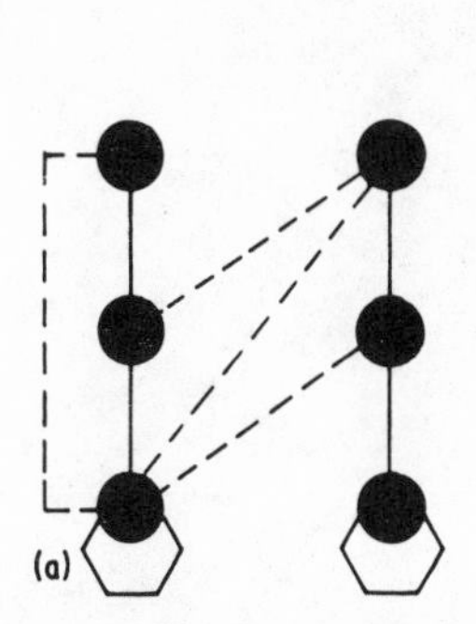

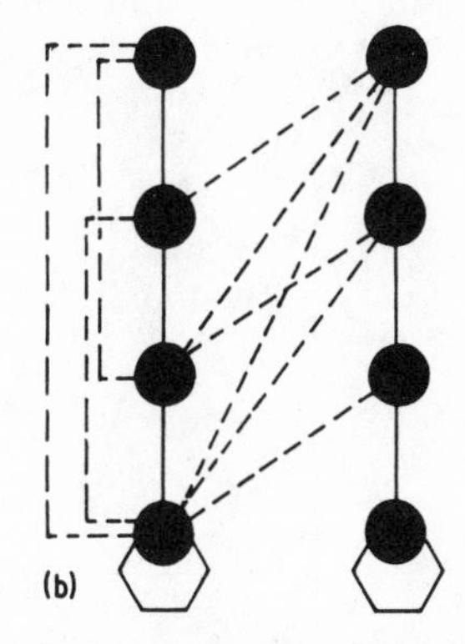

Figure 4.6 Digraphs of six- and eight-species models whose stability is discussed in the text. All analyses have interactions indicated by the full lines. Broken lines indicate interactions that are present in some models and, for simplicity, only interactions from species in the left-hand chains are shown; in the analyses, interactions from both chains were possible. With the exception of intraspecies interactions at the base of the web all lines indicate paired interactions: positive (effects of prey on the predator's growth rate) and negative (effect of predator on the prey's growth rate).

models. For the six-species models without any additional interactions there are $6!/4!$ ways of organizing the two negative terms on the diagonals and $30!/22!$ ways of organizing the eight off-diagonal terms. The numbering of the species is arbitrary, so I must divide the possible number of matrices by 6! to get: $30!/(22! \times 4!) \simeq 10^{10}$ different combinations. Of this vast number I considered only one, though clearly there is a small number of others that would be biologically acceptable.

The magnitude of the interactions was also different from those used in the studies described so far. (I shall return to this in Chapter 7.) Suffice it to say that the predator–prey interactions were chosen to reflect the greater *per capita* effect that a predator has on its prey rather than vice versa.

The results are shown in Fig. 4.7. They are consistent with the results embodied by Equations (4.2a, b) but only qualitatively: (i) As connectance increases there is a sharp transition between models that are likely to be stable and those that are not. (ii) The critical values over which this change takes place are lower in the models with more species. I have not yet determined whether or not these non-random systems give a hyperbolic relationship between n and C, as the boundary between stable and unstable models, suggested by Equations (4.2a, b) for randomly constructed models.

Quantitatively, the results confirm those obtained by Lawlor, in that reasonable models are more likely to be stable than their totally random counterparts. In the six-species models, $s = 2.24$. The average diagonal term was $-1/6$ because two elements were uniformly distributed on the interval $(-1, 0)$. In Equations (4.2a, b), the right-hand side of the inequality is the

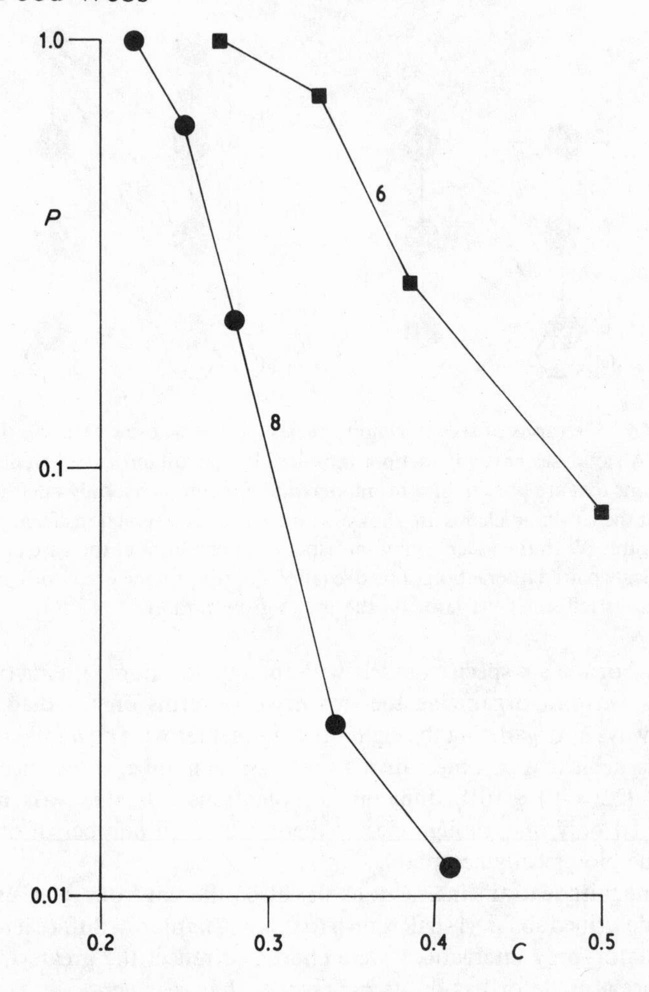

Figure 4.7 The probability (P) that a model of given connectance (C) will be stable, using the six- and eight-species models shown in Fig. 4.6. Note the logarithmic scale of the ordinate. (After Pimm, 1979b.)

average diagonal term (which is constant, and equal to -1, for all species in May's analysis). Substituting these values, I calculated the critical connectance as 6^{-5}, or about 0.001. This value, of course, is for a random symmetrical matrix; the critical value for the actual, ecologically reasonable matrix was about 0.3 (see Fig. 4.7).

It might be concluded that because reasonable models have such improved chances of being stable, the reasonable models observed in the real world are a consequence of the unreasonable ones having been eliminated. This may be so, but there are many reasons why these unreasonable structures (such as loops) are biologically impossible. Although I shall argue that stability places limitations on possible food web structures, this argument seems a trivial explanation for these constraints of 'reasonableness'.

4.2.4 Self-limiting terms and the effects of predators

Perhaps the most contentious assumption involves the diagonal elements of the Jacobian matrix. Stability critically depends upon these, as can be seen in Fig. 4.2: the more negative the diagonal elements, the more likely the system will be stable (see also Yodzis, 1981). Many authors have used the assumption that all the diagonal elements should be negative and of large absolute magnitude. It is a view to which I strenuously object.

Rejmanek & Stary (1979) have drawn attention to Tanner's (1966) work as putative evidence that most populations show self-limiting interactions. This work was discussed in Chapter 1 when I addressed the question of whether natural populations were stable. Negative correlations between growth rates and population densities predominated in Tanner's study: when populations were at high levels, they declined and vice versa. Now this certainly can be achieved by making the a_{ii} in the Lotka–Volterra models negative, but this is not the *only* way in which this result can be obtained. Consider the two-species predator–prey model discussed earlier (e.g. Equation (2.7)). The diagonal element in the Jacobian matrix, corresponding to the predator, is zero (Equations (2.29), (2B.2)). Yet the predator will still show apparently stabilizing changes in density. Simulation of the equations confirms the intuition that if the predator increases, the prey will decrease and the shortage of food will eventually cause a reduction in the density of the predator. The predator's population will behave like those in Tanner's study because the predator is limited by its food supply (prey) and that species, in turn, is limited by some other factors. In contrast, the negative term on the diagonal implies some factor acting to regulate a species' density independently of the species' food supply. This is easy to accept for species at the base of the food chain. A plant may be limited by space, light, water or nutrients, but these factors are not explicitly modelled by the equations. For animals, certain behavioural phenomena can lead to negative diagonal terms of relatively small magnitude. Indeed some negative diagonal elements among the animal species may be necessary for the stability of real systems (Yodzis, 1981). But what if the self-limitation is strong, brought about perhaps by territoriality or a pecking order? By 'strong' I mean a situation that produces a species' equilibrium well below that imposed by its food supply and predators. This is exactly what Wynne-Edwards (1962) envisaged, and it is generally agreed that it requires group selection to achieve it (Lack, 1966, pp. 299–312).

There are some situations which can generate strongly negative diagonal elements in the Jacobian matrix without invoking the pitfalls of group selection. These cases involve the donor-controlled dynamics discussed in Chapter 2.

The Jacobian matrix of the Lotka–Volterra predator–prey model of Equation (2.7) has the sign structure

	(1)	(2)
(1)	−	−
(2)	+	0 .

This is derived using Equations (2.29) and a worked example is shown in Appendix 2B, where X_1 is the prey. The Jacobian matrix for the equivalent donor-controlled systems (Equation (2.9)) is different. Linearizing this system, and using Taylor's expansion, yields

	(1)	(2)
(1)	−	0
(2)	+	− .

As before, the prey are limited by some external factor. The predators gain from the prey by eating dead or doomed animals. The predator's growth rate decreases as its density increases for the same reason that the prey's growth rate did, namely, there are more predators among which to divide the available resources. In such models, every diagonal element will be negative, but these models are not of the same form as those of Gardner & Ashby (1970). The predators do not affect the growth rate of the prey and so there are no *off-diagonal* elements that are negative.

Donor-controlled models provide a conspicuous exception to the results developed so far. Donor-controlled models are usually stable and their local stability is independent of, or even increases with, added complexity (DeAngelis, 1975). Consider the donor-controlled equivalent of Fig. 2.3(b). The Lotka–Volterra model is not qualitatively stable, in contrast to that of Fig. 4.8. Moreover, removal of the predator–prey interaction between X_1 and X_3 (reducing the complexity) increases the stability of system 2.3(b) – the system becomes qualitatively stable. In Fig. 4.8 removal leaves the system unchanged – it remains qualitatively stable.

The reader may object to the 'either' Lotka–Volterra 'or' donor-controlled formats that I have presented. Surely, in the real world, predators will range from having little impact on their prey's growth rates to having large impacts? Certainly, the real world will not be as clear-cut as I have portrayed it. DeAngelis (1975) and May (1979) have shown that intermediate models have intermediate dynamics. The more important is donor control, the more likely that stability will be independent of, or even increase with, complexity.

In summary, many criticisms have been leveled at the result whose essence is expressed by Equations (4.2a, b), namely, model stability is decreased by increasing numbers of species, connectance and interaction strength. The only criticism known to reverse this result is that it does not hold for donor-controlled models. Clearly, a final decision on the relationship of complexity to

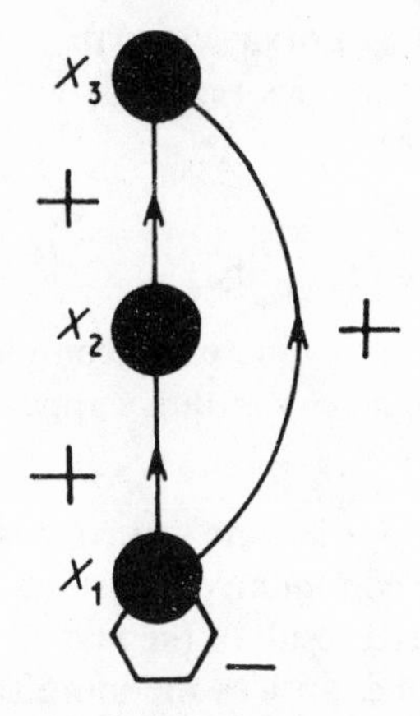

Figure 4.8 A digraph of a donor-controlled model (compare with that of Fig. 2.3(b), where predators do affect the growth rate of their prey). This model is qualitatively stable.

stability must depend on an evaluation of whether donor-controlled or Lotka–Volterra dynamics are the better description of the real world. I shall consider the reponses of donor-controlled dynamics to the effects of large perturbations in the next section and then evaluate in Chapter 5 the implications of the two model formats in light of available field data.

4.3 COMPLEXITY AND STABILITY UNDER LARGE PERTURBATIONS

4.3.1 A preview

The discussion so far has concerned small perturbations, for which there are useful theorems about the eigenvalues which result from linearizing the equations. But what if species densities are subjected to much larger perturbations? Following the removal of a particular species, how does the chance of a food web losing one or more species change with model complexity? I shall show that, again, the answer depends on whether Lotka–Volterra or donor-controlled dynamics are the better description of the real world. With the latter models, increased complexity gives increased stability; with the former, the reverse is true. Both hypotheses have testable consequences and I shall present data which reject donor-controlled dynamics and, with them, one of the otherwise more plausible model systems where complexity does give increased stability. Moreover, there is a bonus: data in Chapter 5 indicate that most natural systems do not withstand perturbations as large as species deletions. In future chapters, these results provide a justification for concentrating on small perturbations (because systems do not withstand large ones) and Lotka–Volterra dynamics (because donor-controlled dynamics permit species removals without further species losses and the real world does not).

4.3.2 Species deletion stability

MacArthur's (1955) assertion that increased complexity dampens the effect of abnormal abundances has a strong intuitive appeal. He provided a definition of stability:

> 'Suppose, for some reason, that one species has an abnormal abundance, then we shall say that the community is unstable if the other species change markedly in abundance as a result of the first. The less effect this abnormal abundance has on the other species the more stable the community.'

The correlate of stability, complexity, was defined as: 'the amount of choice of the energy in going through the web.' A species can attain abnormal abundance in two ways: it can either be above or below its usual equilibrium density. I suspect that, in nature, the greatest deviation will occur when a species becomes extinct (perhaps only locally) through emigration, epidemics or increased vulnerability to predators. Species above equilibrium density will likely encounter food shortages and die off quickly. Unusually high abundances are likely to be of short duration. Consequently, I shall only consider the effects of removing a species from a system.

The definition of species deletion stability is an index, S, on the interval (0, 1) (Pimm, 1979a):

$$S = \sum_{i=1}^{n} r_i p_i, \tag{4.3}$$

where n is the number of species in the community, r_i is a weighting term, the probability that if a species is lost from the community it will be the ith species (hence, the r_i sum to unity). The p_i are the probabilities that, if the ith species is lost, there will be no further losses from the community ($p_i \leqslant 1$).

Can the variation of species deletion stability be predicted? Consider the example in Fig. 4.9. In (a), the six-species system has a simple structure. Only two top-predators X_3 and X_6 can be removed without further species losses, for removal of any other species deprives another species of its sole food supply. In (b), there is greater complexity and any of the six species can be removed. For the simple model (a), modelled by Lotka–Volterra equations, if the complete model is stable, then the model, less either of the top-predators, is also stable (Pimm, 1979a). For models of greater complexity, this is not true. Whether (a) or (b) is more species deletion stable depends on the magnitude of the p_i.

Stability also depends on which species are most likely to be deleted (the r_i). If top-predators are the only species to be lost then (a) will have maximum

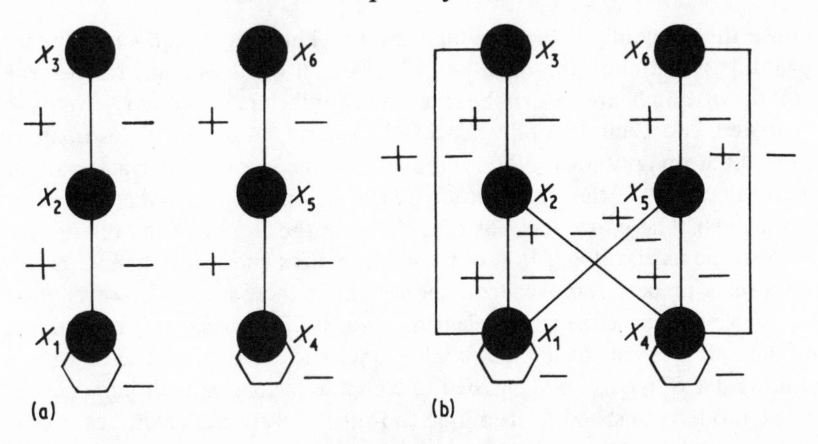

Figure 4.9 Digraphs of two six-species models that differ in species deletion stability (from Pimm, 1979a). For simplicity pairs of predator–prey interactions are indicated with only one line but two signs (+ for the effect of the prey on the predator and – for the reverse).

species deletion stability. If only species at the base of the web are removed then (a) will have no resistance to species deletion because some of the remaining species will lack a food supply. Exactly which species are most likely to be removed from a web by natural causes, I do not know. Thus, the form of the r_i is not known. I shall adopt three equally improbable forms that, hopefully, span all the possibilities.

To investigate species deletion I created samples of systems with population growth modelled by Lotka–Volterra equations (Pimm, 1979a). From models that were both stable and had all species equilibria positive (feasible), each species was deleted in turn. The resulting models were examined to see if the equilibrium was locally stable, had all species with a food supply and had feasible new equilibrium densities.

The values of the parameters for the species not deleted were the same in both the complete model and in the model with the species deleted. This is, perhaps, the least realistic assumption in these analyses. Predator switching, for example, involves interaction terms whose *per capita* effects are functions of the prey densities instead of the constant values assumed here. With local stability analyses, this does not constitute a major source of error, since, by definition, species densities are approximately constant. With large perturbations, however, changes in the magnitude of the interaction coefficients may be large. These analyses must be considered only a beginning.

(a) Simulation details

The analyses involved choosing both the a_{ij} and b_i terms in the Lotka–Volterra models. In some of the studies described so far, it has only been necessary to choose the elements of the Jacobian matrix. This has a number of distinct advantages over choosing the a_{ij} and b_i. First, it involves fewer calculations, for if the a_{ij} and b_i are chosen the resulting equilibrium densities (X^*) must be calculated and their feasibility checked. Second, it is easy to estimate an approximate magnitude for the components of the Jacobian matrix (recall that these are the X^*, the equilibrium densities, and the a_{ij}, the interaction coefficients). The b_i are less obvious: they are the slopes of the relationship between the natural logarithm of population sizes and time when all but the one species has been removed from the system. In these studies, however, it was impossible to choose the matrix elements directly. There is no way of determining the new elements of the matrix after one species has been deleted without a knowledge of the a_{ij} and b_i needed to compute the new equilibrium densities.

The models consisted of from one to four three-species chains, each chain consisting of a plant, a herbivore and a carnivore (Fig. 4.9). Up to five predator–prey interactions were added to increase the complexity of the models. These interactions could be within chains (which involved species feeding on more than one trophic level) or between chains. For plants that increase at low densities when herbivores are absent, $b_i > 0$. For animals, which would decline in the absence of their prey, $b_i < 0$. Only species at the base of the chains were self-limited ($a_{ii} < 0$); elsewhere $a_{ii} = 0$.

In the models $a_{ij} = a_{ji} = 0, j \neq i$, unless i was a predator on j. In the case when i did prey upon j, then

$$\begin{aligned} 0 \leqslant a_{ij} \leqslant 0.1 \\ -1 \leqslant a_{ji} \leqslant 0. \end{aligned} \tag{4.4}$$

The exact values were chosen randomly with a uniform distribution along the intervals specified by Equation (4.4). I shall discuss the choice of parameters in Chapter 7. For now, note that the choice of these intervals implies that the effect of a predator population on a prey population is of greater magnitude than the effect of the prey on the predator.

The choice of the b_i is difficult not only because of their ecological intractability, but also because the condition of feasibility imposes severe constraints on their magnitudes. I fixed each b_i for a basal species at 1.0 and selected a_{ii} for basal species over the interval $(-1, 0)$. Minus the ratio of these quantities gives the equilibrium density of the basal species in the absence of any predation (its carrying capacity). This choice of parameters scales the system, since the median carrying capacity would be 2 and less than 1.0% of the models would have carrying capacities greater than 100. In short, the X_i are densities per unit of area chosen to give a median carrying capacity of 2.

Consideration of what magnitudes the other b_i should attain suggests that they should be small to ensure that the predators have feasible equilibrium densities. I chose b_i to be -0.02 for species not at the base of the web. Although the b_i are fixed, the elements in the Jacobian matrix and the X^* are still random variables as a result of the random selection of the a_{ij}.

For each model structure, I produced at least 1000 random webs, calculated the X^* and checked them for feasibility. For those that passed this test, the Jacobian matrices were calculated and their eigenvalues extracted numerically. Only those systems that were locally stable were used as the basis for the study. From this set, each species was deleted in turn and the resulting $n-1$ species system tested to see whether:

(i) Each predator had at least one prey species.
(ii) The new system was not singular (this condition is discussed on p. 32 and, most often, it involved n species of predator sharing exactly the same $n-1$ species of prey).
(iii) The new system was feasible.
(iv) The new system was locally stable.

Systems passing all the above tests were deemed species deletion stable.

(b) Effects of species deletion

I shall discuss two kinds of results: (i) how species deletion stability varies between webs of differing numbers of species and complexities; and (ii) how species deletion stability varies between species within a web.

(i) Figure 4.10 shows species deletion stability averaged over all the species in a web against the connectance for webs with differing numbers of species. The average species deletion stability decreases with increasing connectance. For a given connectance, it also decreases with increasing species number.

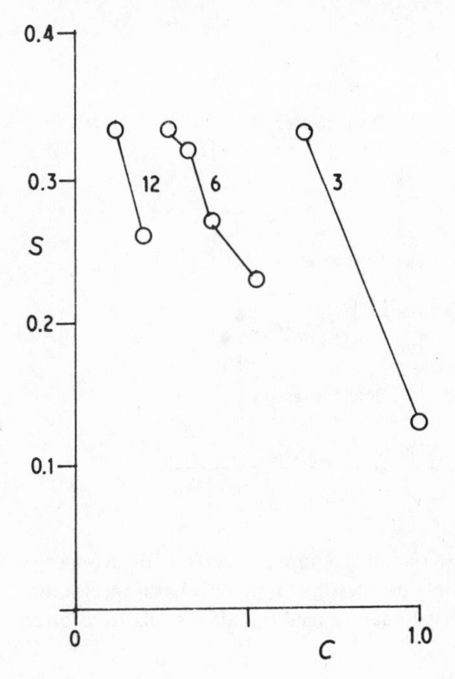

Figure 4.10 Species deletion stability (S) against connectance (C) for models with differing numbers of species (From Pimm, 1980c.)

Only a limited portion of the range of connectances can be shown for each species number. The limits are imposed by two factors. The models are three-, six- and twelve-species systems organized into chains of three species, for example, a plant, its herbivore and a carnivore. The simplest models involve interactions only within food chains. The corresponding connectances are 2/3, 4/15 and 8/33 for the three-, six- and twelve-species models respectively. In all of these models, only the carnivore in each chain can be removed without any further losses of species. The upper bound for species deletion stability is therefore 1/3. The lower bound is imposed by the difficulty of finding sufficient feasible and stable systems to examine for the effects of species deletion (Pimm, 1980c).

(ii) Which species can be removed without further losses of species from the web? The results for six-species models show that species deletion stability averaged over all species changes only slightly with increasing complexity. There are, however, marked differences depending on which species are deleted from the web (Fig. 4.11). For species at the base of the web, increasing complexity *does* increase species deletion stability. But, increasing complexity rapidly decreases the chances that a model food web will persist in the absence of its top-predators. MacArthur's (and others') intuition about stability was correct, in part, but not complete. The argument he made considered the

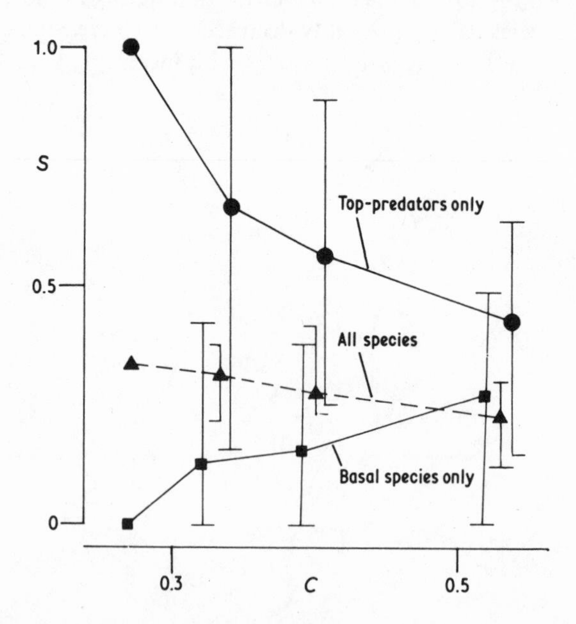

Figure 4.11 Species deletion stability (S) against connectance (C) for six-species models. Results are for systems where (i) only top-predators, (ii) only basal species, and (iii) all species are deleted in turn. Intervals are ranges for all models of a given connectance. (From Pimm, 1979a.)

effects of perturbing a system 'from below' and is correct for such perturbations. Only when perturbations 'from above' are included are the results reversed.

To explore this further, I examined a set of ten webs, each with twelve species and with five additional predator–prey interactions over and above the minimum implied by the four three-species chains. These models have a connectance of 0.2. Figure 4.12 shows an example with each species coded according to the proportion of removals that caused further species losses. The additional predator–prey interactions were placed randomly. From the set of ten webs, I obtained the results shown in Fig. 4.13:

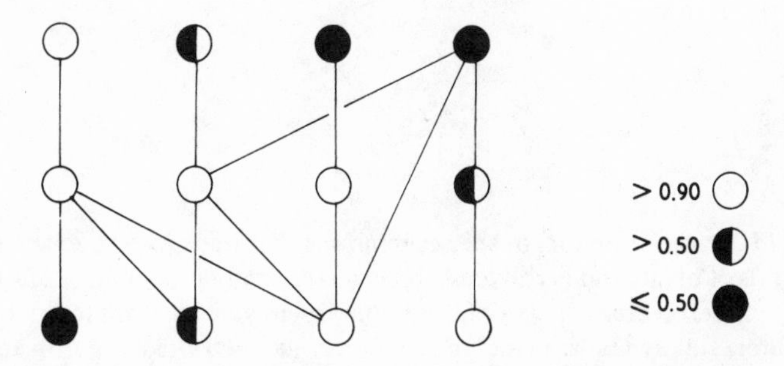

Figure 4.12 Examples of a twelve-species model to illustrate differences in the effect of removing individual species from model webs. Legend: proportion of removals that leads to further species losses. (From Pimm, 1980c.)

(a) Plant-species removals cause least loss when the herbivore feeds on a variety of other species, that is, it is polyphagous.

(b) Removal of herbivores shows a comparable pattern. The more polyphagous are the herbivore's predators the less the effect of removing the herbivore.

(c) Predator removal shows that further species losses are more common when the herbivores on which the predators feed are polyphagous. This effect is in the *opposite* direction to those of plant and herbivore removals. With plant and herbivore removals, the more complex is the web the more species deletion stable it will be. For plants and herbivores, increased complexity does give increased species deletion stability. The effect of removing the predators, however, is sufficiently strong to counter these effects, producing a pattern of species deletion stability that decreases with increased complexity.

Some factors appear not to affect species deletion stability. In particular, whether the predator is monophagous or polyphagous has no bearing on the effects of its removal (Pimm, 1980c).

These results suggest the question: is it possible to develop a reasonable model in which the effects of carnivore removal are minimized or eliminated

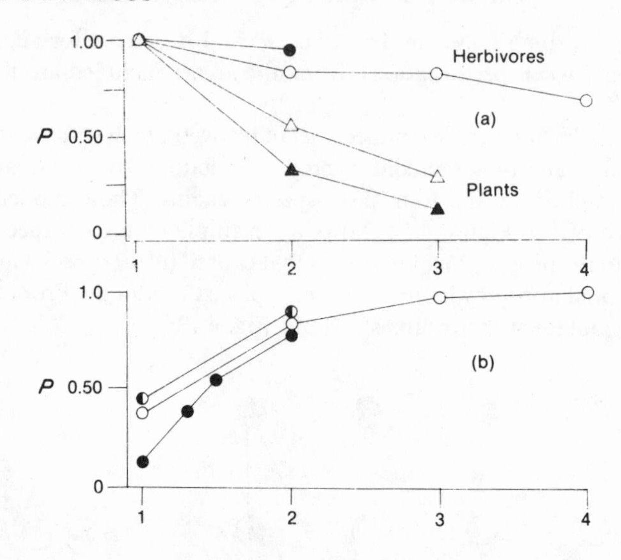

Figure 4.13 Proportion (*P*) of species removals that leads to further species losses. (a) Removals of plants and herbivores; abscissa is the minimum number of species that the plant's (or herbivore's) predator feeds upon. Open symbols: plant (or herbivore) has one predator; full symbols: plant (or herbivore) has two predators. (b) removal of predators; abscissa is the mean number of prey the herbivore(s) (upon which the predator removed feeds) feeds upon. Open symbols: predator has one prey; half-open symbols: predator has two prey; full symbols: predator has three prey. (From Pimm, 1980c.)

while those of plant and herbivore removal are not, so that as the complexity of the web increases so does its stability?

To answer this question, consider which species become extinct following a species' removal. The answer is usually, but not always, clear. When plants or herbivores are removed, it is the species which feed upon the deleted species that are lost. When a carnivore is removed the answer is less obvious. Figure 4.14 shows two webs, one with a monophagous herbivore (a) and the other with a polyphagous herbivore (b). In Fig. 4.14(a), the removal of the carnivore will not cause any further species losses (Pimm, 1979a), while in Fig. 4.14(b), removal of the carnivore will almost certainly cause some species losses within the plant trophic level. With Lotka–Volterra dynamics, if the same carnivore parameters are used in (a) and (b) then the herbivore's equilibrium density will be the same regardless of its population parameters or the number of plant species on which it feeds. In the absence of the carnivore, herbivore equilibrium densities will be very different from (a) to (b). Specifically, the more plants on which the herbivore feeds, the higher will be its density. The plants' densities depend, in part, on the herbivore's density: the higher the latter, the lower will be the former. Thus, each plant species has a greater chance of

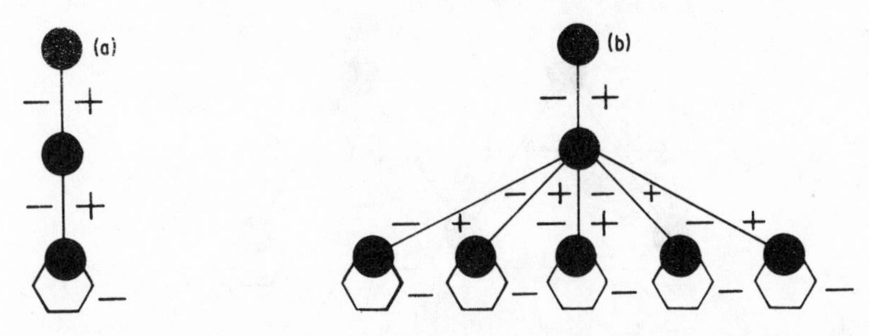

Figure 4.14 Two digraphs. In (a) removal of carnivore will not cause loss of plant species; in (b) removal of carnivore will almost certainly cause loss of one or more plant species. Single lines are used to represent the pairwise predator–prey interactions but both + and − effects are indicated. (After Pimm, 1980c).

extinction the more plant species there are. The chance of at least one plant species becoming extinct would increase with increasing numbers of plant species, even if the individual probabilities of extinction remained the same. Simply, the chances of at least one plant species being lost increase rapidly with the herbivore's increasing polyphagy. Holt (1977) discusses this in more detail for a wide variety of models. Actually, a plant does not need to be lost for species losses to occur. Herbivores feeding on the plants (but not shown in the figure) may become extinct because the equilibrium densities of their food supply become too low to support them.

Now, it is possible for a predator to feed upon a prey and not affect (least of all, uniquely determine) the prey's equilibrium density. Donor-controlled dynamics have this characteristic. Increased donor control increases the chances of local stability (DeAngelis, 1975, May, 1979), and with donor control, increased complexity also gives increased resistance to species deletion. Shown in Fig. 4.15 are three prey species (X_1, X_2, X_3) fed upon by two predators (X_4, X_5). The links marked α and $1-\alpha$ may be zero, in which case the model has lower connectance. (These terms represent the proportion of X_2 that is consumed by each of its two predators.) If this system were modelled by Lotka–Volterra dynamics, deletion of any of the three prey species might cause loss even if α or $1-\alpha$ were not zero. But when, for example, $\alpha = 0$, the deletion of X_1 inevitably causes the loss of X_4. None the less, this tendency for added complexity to increase stability is more than countered by the decreased resistance to predator removal. For example, when $\alpha = 0$, removal of X_4 or X_5 can have no effect on the prey. When $\alpha \neq 0$ removal of X_4 can cause X_3 to be lost and, similarly, removal of X_5 can cause X_1 to be lost.

With donor-controlled dynamics, the results are different. Removal of the predators has no effect on the prey. Deletion of a prey species will only result in the loss of a predator if that predator feeds only on the prey removed. Consequently, increased complexity ($\alpha \neq 0$) gives increased species deletion stability when averaged over all five species.

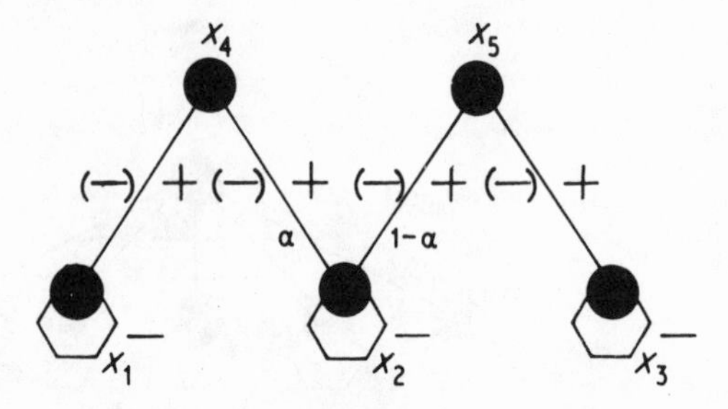

Figure 4.15 A five-species digraph, where α and $1-\alpha$ represent the proportions of the biomass of X_2 consumed by X_4 and X_5 respectively. Under donor-controlled dynamics signs marked (−) are zero: the predators do not effect the growth rate of their prey and increased complexity (α, $1-\alpha$, are not zero) gives increased species deletion stability. Under Lotka–Volterra dynamics signs marked (−) are negative and increased complexity decreases species deletion stability. The + signs indicate that prey always have a positive effect on the growth rate of their predators. (From Pimm, 1980c.)

4.4 SUMMARY OF THEORETICAL RESULTS

There are three principal results:

(i) Most local stability analyses suggest that increased complexity, increased species numbers and increased interaction strength should all lessen the chances that a system will be stable.

(ii) Equations (4.2a, b) are derived for randomly organized systems. They suggest that stable systems will be bounded by a hyperbolic relationship between the connectance (C) and the number of species (n) in the system, provided the average interaction strength is constant. The hyperbolic relationship has not been obtained for models that are biologically reasonable. Still, the results that have been obtained are at least qualitatively consistent with it.

(iii) For larger perturbations (species removals), stability also declines with increasing complexity but, more critically, depends on which species are removed. When species at the base of the food web are removed, greater complexity leads to lessened chances of further species losses. When species at the top of the food web are removed the opposite is true and, overall, it is this effect that predominates in Lotka–Volterra models.

The models that produce these results have assumptions that have been questioned. Yet, only one assumption reverses the conclusion. With donor-controlled models, increased complexity either increases or does not affect the chances of local stability and, similarly, increased complexity increases the chances that no species will be lost following a species removal.

These results suggest ways of resolving the question: does stability place upper or lower bounds on food web complexity? The first result suggests that systems differing in either species numbers, connectance or interaction strength ought to differ in stability, provided the other two variables could be assumed constant. However, recall the argument in the Introduction. The systems observed are those that have survived perturbations. It might be surprising if natural systems differed greatly in their response to small, experimental perturbations. The second result seems more promising. It suggests there will be a precise functional form for the bound imposed on connectance by stability: it should decline in a hyperbolic fashion with increasing species number or interaction strength.

The third result suggests an indirect method of investigating the relationship between stability and complexity by testing whether donor-controlled or Lotka–Volterra dynamics are better at describing interactions in the real world. It is a suggestion that requires elaboration. There are both quantitative and qualitative differences between the predicted effects of predator removal in Lotka–Volterra and donor-controlled models. The differences are (i) with Lotka–Volterra dynamics, the removal of the predator should alter the equilibrium density of the prey; with donor-controlled dynamics it should not. This result was derived earlier (Equation (2.11)). (ii) Except for the simplest models (e.g. Fig. 4.14(a)), some species will generally be lost from the system following predator removal when the Lotka–Volterra assumptions obtain. There will not be species losses following predator removal with donor-controlled dynamics by definition. These predictions are testable. And, if donor-controlled dynamics are rejected, then so is the most biologically plausible model system where increased complexity gives increased stability.

5 Food web complexity II: empirical results

The last chapter concerned the question: does the requirement of stability impose upper or lower bounds on how complex a food web can be? This chapter presents three kinds of empirical evidence which bear upon this question. The first is direct tests of the relationship between complexity and stability. The second involves attempts to demonstrate the theoretically suggested hyperbolic relationship between the connectance and the number of species. Finally, there is the indirect evidence on the consequences of species removals; this evaluates whether donor-controlled dynamics are good descriptions of the real world.

5.1 DIRECT TESTS

At least conceptually, the recipe for testing the complexity–stability hypothesis ought to be easy. Take a couple of dozen ecosystems with food webs of varying complexity, number of species and interaction strength then scramble with a standardized perturbation. Follow closely until species densities have returned to equilibrium or not, as the case may be, and digest the conclusions.

The first problem with this recipe is that the systems we observe are those that have resisted perturbations. They might be considered stable, in at least one of the many definitions of that word, merely as a consequence of their existence. Consequently, tests must involve some relative measure of stability. Yet, there is no guarantee that relative measures of stability, for example return times, should vary in the same way with complexity as such absolute measures as whether or not the population densities return to equilibrium. The second problem is that studies have concentrated only on the effect of species number on stability.

5.1.1 Species number and stability

Several experimental studies have concentrated on the relationship between species number and stability. If this approach is to be related to the theoretical studies of Chapter 4, then it must be assumed that connectance and interaction strength are relatively constant. I shall show later that this assumption is invalid. For now, the relationship between species number and how systems respond to a mixed bag of perturbations raises some interesting questions. In its own right, it is an interesting problem.

(a) McNaughton's results

(i) *Grassland plants*

McNaughton (1977) discussed two experiments on plants. In one (Hurd, Mellinger, Wolf & McNaughton, 1971, Mellinger & McNaughton, 1975), systems were perturbed by adding a nutrient to the soil. In the second (McNaughton, 1977), the perturbation was obtained by the addition of a grazing herbivore. The first study used two abandoned hayfields in central New York State. One field had a greater number of plant species than the other (about 30, compared to 20, per 0.5 m^2 plot). Plots in each field were either fertilized (the perturbed plots) or left alone (the control plots) during the spring. In the summer of the same year, plant samples were taken in each field from each of the control and perturbed plots.

Several assumptions must be made beyond those of constant interaction strength and connectance. First, the perturbations were not to species densities, nor did they involve species removals. The nutrient perturbation was probably used up by the system over some intermediate period of time. Second, the measure of stability was not whether equilibrium was recovered or not, nor how quickly the systems returned to equilibrium. Indeed, all natural communities might be expected to survive perturbations of this sort if variations in nutrient supply were common in the real world. Rather, the measure was how similar in plant composition were the controlled and perturbed plots. This may reflect how quickly the populations returned to equilibrium, but it will also reflect differences in how far each system was moved from equilibrium by the perturbation. The similarity between control and perturbed plots was measured by two indices: (i) how many species were present in the 0.5 m^2 plots; and (ii) equitability. Equitability is a measure of evenness in species abundance. For example, a sample of four species has a maximum evenness if these species have relative abundances of 0.25 each and less evenness if one species predominates and the others are rare (e.g. 0.97, 0.01, 0.01, 0.01). Table 5.1 shows that the number of species per sample was not statistically different in control versus perturbed plots in either the species-poor or species-rich fields. Yet, there were significant changes in the equitabilities in the species-rich field but not in the field with fewer species. The equitabilities represent changes in species abundance. That these changes were greater in the system with more species seems to substantiate the theory that systems with more species are less stable.

In the second plant study McNaughton (1977) compared two sites that differed in diversity. *Diversity* has a technical meaning: it is an index that combines both equitability and species number (Table 5.2). The perturbation was obtained by the presence of an African Buffalo (*Syncercus caffer*). The more diverse site showed significant differences between grazed and ungrazed plots, while the less diverse site did not. These results are similar to the first study and the conclusions the same.

Table 5.1 Plant community characteristics with and without a perturbation from added fertilizer (from McNaughton, 1977).

	Control plots	*Perturbed plots*	*Statistical significance*
Number of species per 0.5 m^2			
species-poor field	20.8	22.5	ns†
species-rich field	31.0	30.8	ns
Equitability			
species-poor field	0.660	0.615	ns
species-rich field	0.793	0.740	< 0.05
Primary productivity ($g\ m^{-2} day^{-1}$)			
species-poor field	28.0	42.0	< 0.05
species-rich field	19.7	22.8	ns

† ns = not significant.

Table 5.2 Plant community characteristics with and without a perturbation from grazing by an African Buffalo (from McNaughton, 1977).

*Diversity**	*Control plots*	*Perturbed plots*	*Statistical significance*
More diverse stand	1.783	1.302	< 0.005
Less diverse stand	1.069	1.357	ns

Live, standing crop biomass reduction (perturbed, as a percentage of control) was 11.3% for the more diverse stand and 69.3% for the less diverse stand

* A technical measure that combines both the number of species and the evenness of their abundances.

McNaughton trenchantly draws the exact opposite conclusion and his arguments clearly illustrate the diverse meanings of 'stability'. In the first study the amount of plant material produced during the growing season was different in control and treated plots in the species-poor field but nct in the species-rich field (Table 5.1). In the second study the buffalo's grazing reduced the proportion of living plant material by 69% in the less diverse stand, but only by 11% in the more diverse stand (Table 5.2). If stability is measured in terms of changes to the amount of plant material, then certainly the more diverse communities are the more stable. These diametrically opposite conclusions are related. McNaughton argued that the lack of change in plant production with added nutrients in the species-rich field was because there were more species. With more species there was a greater potential for at least one species to change dramatically in density when the system was perturbed, and so accomodate the change in nutrient availability. The mechanism is more obvious in the second example. Here, when the systems were grazed by the herbivore, the more diverse site had the greater potential for at least one species to change in abundance to compensate for the grazing. The species that did this was identifiable in the study. In contrast, the less diverse site did not possess a species that allowed it to compensate for the effects of grazing.

It might be concluded that whether stability increases or decreases with species number depends solely on definition. McNaughton's comments are more important and more subtle than this. Diversity of plant species is a structural property of an ecosystem, while changes in plant production are a functional property. That the function depends on the structure suggests a broad significance to a knowledge of food web properties. Simply, McNaughton's results suggest inter-relationships between structure and function and also suggest that species number, alone, may be important in affecting an ecosystem's functional response to perturbations. These are topics to which I shall return in Chapter 10.

(ii) *Insects*

The first of two insect studies is the insect sampling that accompanied the first plant study discussed above (Hurd *et al.* 1971). Early and late in the growing season, Hurd *et al.* measured the numbers of both herbivorous and carnivorous insects in their plots (Table 5.3). The field with more plant species also had more insect species. For the herbivorous insects, the changes in diversity between controlled and perturbed plots were greater in the species-rich field early in the season (when the diversity increased). This was not true later in the season (when the diversity decreased). The diversity of carnivorous species increased with the perturbation more in the species-rich than the species-poor field early in the season. Late in the season, the diversity of species in both fields increased with the perturbation but not by amounts that differed significantly between treatments.

These results are mixed. Only in half of the cases are the results in the same direction as those for the plants where, following a perturbation, the changes in diversity were greatest in the species-rich field.

(b) Pest outbreaks

Pimentel (1961) took a very different, though still experimental, approach. His study tested the effects of increased plant and insect diversity on the probability that there would be a pest outbreak on the cultivated plant *Brassica oleracea.* The plants were grown in single-species plots and alongside many other plant species in a 15-year-old field. In the old field there were fewer taxa associated with *Brassica* (27 insect taxa, versus 50 in the single-species plot), but there were an additional 300 species of plants with perhaps 3000 species of animals to provide the extra diversity.

Aphids, flea-beetles and lepidopteran populations reached outbreak levels in the single-species planting, but populations were much lower in the mixed-

Table 5.3 Animal community characteristics with and without a perturbation to the plants of added fertilizer (from Hurd *et al.* 1971).

	Control plots	*Perturbed plots*	*Statistical significance**
Herbivore diversity†			
Early in the season			
species-poor field	3.40	4.20	< 0.01
species-rich field	3.65	5.50	
Late in the season			
species-poor field	3.35	5.45	< 0.01
species-rich field	5.00	4.45	
Carnivore diversity			
Early in the season			
species-poor field	1.85	2.35	< 0.05
species-rich field	1.75	3.05	
Late in the season			
species-poor field	1.5	2.5	ns
species-rich field	1.3	2.05	

* Tests whether there are differences in the changes between control and perturbed plots given that there are already differences between the species-poor and species-rich fields. Technically, the value is the significance of the interactive effects given that purely additive effects are already included in the statistical model.

† Diversity is an index that combines both the numbers of species and how evenly their abundances are distributed.

species planting. The former system was considered less stable than the latter. Pimentel concluded that plant and/or insect diversity played a role in preventing population outbreaks. He also discussed a variety of studies where increased plant diversity appeared to minimize the chances for pest outbreaks. For example, he quoted several studies on various species of pine trees which, when grown in mixed stands with other species, were much less likely to be attacked by the white-pine weevil (*Pissoides strobi*). Though there are many similar examples there are also some exceptions and complications. The bracken fern (*Pteridium aquilinum*) grows over large portions of the English countryside in stands that contain very few other species. The species is common worldwide yet does not appear to suffer insect outbreaks (Lawton, 1974). Similarly, May (1973b) mentions *Spartina*-dominated, salt-marsh ecosystems as a monoculture that appears quite stable. In contrast, the spruce budworm (*Choristoneura fumiferana*) has a highly unstable population (Watt, 1965) and feeds on balsam fir (*Abies balsamea*) and white spruce (*Picea glauca*) which form species-poor stands over much of eastern Canada. This system is only simple floristically, for there is a large variety of parasitic and predatory species which attack the budworm (Morris, 1963).

If there is a consensus in these results I cannot find it. McNaughton's results show that the more diverse systems show greater changes in species abundance when the systems are perturbed. Pimentel's results show that insect herbivores reach levels detrimental to their food supply in species-poor, but not species-rich, systems. And there is anecdotal evidence that some species-poor systems suffer insect outbreaks while others do not. The different results may reflect the differing definitions of stability or genuine differences between ecosystems. They may also indicate that there is no achievable consensus on the relationship between stability and species number. But then the theory suggests that consensus would only be expected if the assumptions about connectance and interaction strength being independent of species number were reasonable ones. In the next section I shall show they are not. If there is a conclusion to be made from these results, I suggest it is that species number alone is no guide to the stability of a system.

5.1.2 The relationship between species number, connectance, and interaction strength

There are several studies of food webs where species number, n, and connectance, C, have been calculated. The interaction strength, s, may be unknown, but assuming it to be constant, though hardly ideal, is one less assumption than was required by the previous studies. If all the food webs are stable, Equations (4.2a, b) predict that their parameters should be bounded by a hyperbolic function, $Cn = \text{constant}$. Several studies, drawn from two sources, support this prediction. The first (Fig. 5.1) is a heterogenous collection of studies assembled, in part, by Cohen (1978), and augmented by a variety of other published results discussed in various parts of this book. Yodzis (1980) has also analysed Cohen's data with comparable conclusions to those presented here. The second source is a study by Rejmanek and Stary (1979) who present data for one group of insects and their prey; all the data were collected in comparable systems by the same authors using similar methods.

All the systems in Fig. 5.1 have approximately the same product of Cn ($= 3.1$). They should differ little in their stability or else the stability should depend critically on the magnitude of the unknown parameter, namely, the interaction strength. I would not necessarily expect the more diverse communities to be more or less stable, as I did in the previous section. Simply, the previous assumption that connectance is independent of the number of species in the web is untenable.

It might seem that the data in Fig. 5.1 are an encouraging conclusion following the considerable theoretical machinations in Chapter 4. I disagree. The hyperbolic relationship between C and n will almost certainly result from a very simple biological assumption that has nothing to do with constraints of stability (Pimm, 1980a). Suppose each species in a community feeds on a number of species of prey that is independent of the total number of species in the community. This seems to be the most parsimonious assumption. The

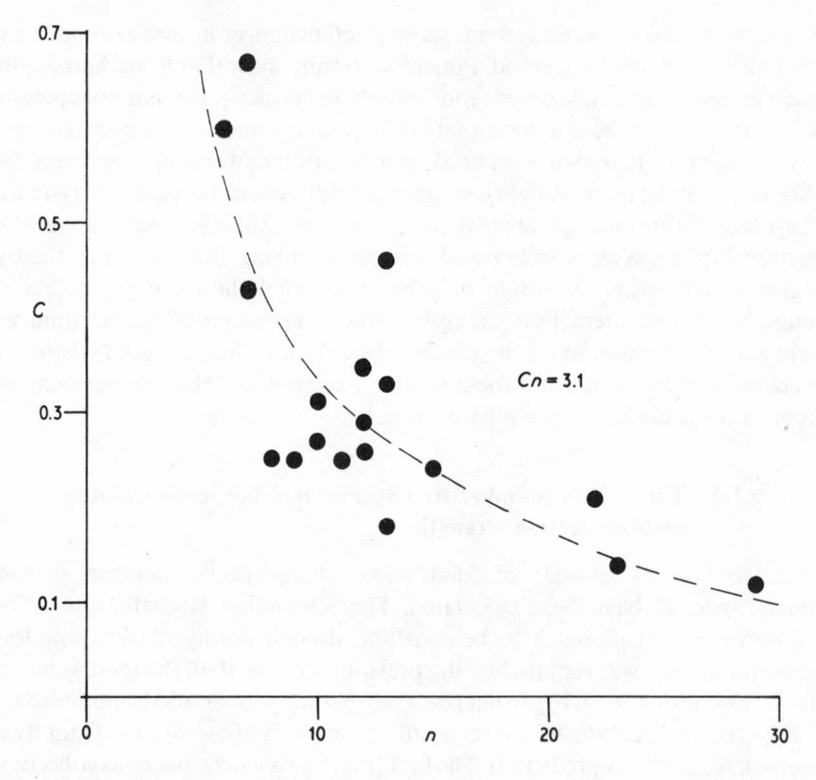

Figure 5.1 Connectance (C) versus species number (n) for a range of food webs. Broken line is the least squares fit. The data are presented in Table 6.1.

alternative of feeding on more species the more there are, conjures up an amazing 'Jack-of-all-trades' with none of the specialized anatomy and behaviour that most species possess. Let the average number of species with which any one species interacts be k, which includes the species' predators as well as its prey. If the number of species in the community is n, the total number of interaction coefficients in the web is kn. Now, there are $n(n-1)$ possible interactions (excluding intraspecific affects). The connectance, C, is

$$kn/[n(n-1)] = k/n-1. \quad (5.1)$$

For large n this will be indistinguishable from Cn = constant and provides a biological reason why C should decrease with n quite independently of any limitations imposed by dynamics. The results of Fig. 5.1 do not test whether stability places restrictions on the complexity of a web (for a given species number) because simple biology will give the same result. But neither do these data reject such a possibility.

In addition, it is possible that there will be changes in the interaction strength as k changes. The fewer species of prey which a predator exploits, the more likely the predator will be able to exploit its prey efficiently and thus exert a larger *per capita* effect on its specific prey population. Thus connectance, for a given number of species, might be negatively correlated with interaction strength. In short, systems with the higher product of Cn might not necessarily be the less stable: interaction strength may be reduced in compensation.

Clearly, if Equations (4.2a, b) are to be tested, a measure of stability is required related to a measure of complexity, suitably corrected for interaction strength and the number of species. This is what McNaughton (1978) attempted using nCs^2 as the corrected measure of complexity. The resistance to the effects of grazing was the measure of stability. Unfortunately the results are clouded by a statistical artifact (Harris, 1979, Lawton & Rallison, 1979). In concept, this study is the most satisfactory attempt to evaluate directly the hypothesis embodied by Equations (4.2a, b) and it is to be hoped that it will be repeated using acceptable statistics.

5.2 INDIRECT TESTS

An indirect line of reasoning is based on the results described earlier involving donor control. With donor-controlled dynamics, the destabilizing effects of the predators on the prey populations are absent by definition. So, under these dynamics, the removal of predatory species from a community should have no effect on either the species densities, or (particularly) the species composition of the remaining system. Removal of prey species is less likely to result in species losses the more complex the web. Combining these results gives an expected positive correlation between stability and complexity. But are donor-controlled dynamics a better description of the real world than those of Lotka and Volterra?

There are many studies where predators have been removed. Those studies assembled in Table 5.4 (Pimm, 1980c) were selected according to several criteria. Foremost was that the food web from which the species was deleted had been described explicitly, or else the study was directly comparable to another study where this was true. This excluded all the terrestrial studies I could find and restricted the sample to studies of the organisms of freshwater ponds and rocky, intertidal habitats. Though Table 5.4 is large and complex its essential feature is simple. Of the 19 studies only two did not show either marked changes in species densities or species losses following the removal of the predator (webs 11 and 12 in Table 5.4). Naturally, negative results will attract less attention and are less likely to be published. None the less, the general conclusion that most systems do not retain all their original species when one species is removed is supported by other authors (Connell, 1975, who reviews a variety of marine, freshwater and terrestrial results; Paine, 1980, who considers marine systems; and Harper, 1969, who discusses herbivores and their effects on plant communities).

Table 5.4 Effects of removing predators (from Pimm, 1980c).

Web Species removed	*How removed*	*Predator's prey*	*Effects of predator removal*	*Source*
1 *Enhydra* (Sea otter)	man; local conditions – cannot feed in deep water water	*Strongylocentrotus*, (Sea-urchin), fish and molluscs	*Strongylocentrotus* increases dramatically	Estes, Smith & Palmisano, 1978
2 *Strongylocentrotus polyacanthus*	*Enhydra*	*Laminaria* and a variety of other macro-algae	*Agrarum* and *Thalassiophylum* (algae) that are resistent to the predator are out-competed by *Laminaria*. Increased competition eliminated various species of red algae and *Alaria* (algae)	Dayton, 1975a
3 *Strongylocentrotus purpuratus*	*Pycnopodia* (starfish), *Anthopleura* (anemone), local conditions – strong wave action, experimentally	Drift and *Hedophyllum* (algae)	With predators, only coralline algae survive. Predator removal allows *Hedophyllum* and the species associated with it to survive	Dayton, 1975b
4 as 3	experimentally	not given, but presumably as 3	*Lithothamnion* (corralline algae) survives in presence of predator, but is out-competed by *Hedophyllum* when predator is removed	Paine & Vadas, 1969; Paine, 1980†
5 *Strongylocentrotus frasiscanni*	experimentally	*Laminaria* is only species listed, but predator can	*Lithothamnion* survives in the presence of the predator but is out-competed by *Laminaria*, and	Paine & Vadas, 1969

		survive in its absence	overgrown by diatoms *Ulva* and *Halosaccion* (red algae) when predators are removed	
6 *Echinometra*, *Diadema*, *Heterocentrotus* (Sea urchins)	local conditions: wave action	algal lawn	With predators corals survive, without them they are out-competed by the algae	Dart, 1972
7 *Paracentrotus* (Sea urchin)	experimentally, local conditions not fully understood	algae	A variety of algal species and the animals dependent on them invade essentially bare areas following predator removal	Kitching & Ebling, 1961
8 *Stichaster* (Starfish)	experimentally	specialized: 76% of the prey are *Perna* (mussel)	'Rapid domination . . . by *Perna* . . . at the expense of the other resident, space-requiring species'	Paine, 1971, 1980†
9 *Acanthaster* (Starfish)	various coral symbionts (*Trapezia*, *Alpheus*) prevent predator from feeding on coral	prefers *Pocillopora* (branching corals), but will take other non-branching species	*Pocillopora* corals persist only in absence of predator, otherwise they are gazed close, or to, extinction	Glynn, 1976
10 *Pisaster* (Starfish)	experimentally	very generalized	Predator removal allows *Mytilus* (mussels) to remove most of the other species in the community through competition or loss of food supply	Paine, 1966
11 *Pisaster*	experimentally	as 10	Predator removal did not lead to increase in *Mytilus*. Larval *Mytilus* were probably scarce in the plankton during this experiment	Connell (in preparation)

Table 5.4 (continued)

Web Species removed	*How removed*	*Predator's prey*	*Effects of predator removal*	*Source*
12 *Katharina* (Chiton)	experimentally	*Hedophyllum* and other algae	Presence or absence does not affect the rate at which the algae recovers when it, too, has been removed	Paine, 1980† Dayton, 1975b
13 *Acmaea* (Limpet)	experimentally	coralline algae	One change within the five species of coralline algae; one species became commoner at the expense of another. Relative abundance of coralline algae versus *Hedophyllum* unchanged	Paine, 1980†
14 *Patella* (Limpet)	experimentally	*Algae*	Dramatic increase in the abundance of the algae	Jones, 1948
15 *Lepomis* (fish)	experimentally, local conditions: dense vegetation	principally larger zooplankton but will take smaller species if larger ones are unavailable	A large species, *Ceriodaphnia* (cladoceran), predominated after predator removal at levels from 44% to 97% depending on nutrient levels in the pond. Smaller species (e.g. *Bosmina*, also a cladoceran) and other herbivores (rotifers and copepods) are eliminated through competition	Hall, Cooper & Werner, 1970
16 Odonates coleopterans and	experimentally	*Daphnia*	Largest species *Daphnia* (cladoceran) increases, smaller	Hall *et al.* 1970

other planktonic predators			species decrease; data are described as 'weak' and one predator, *Chaoborus* (dipteran), was not removed	
17 *Anax* (naid) (benthic predator)	experimentally	mainly *Chironomus* (dipteran), but also *Caenis* (ephemeropteran)	*Chironomus* increases, *Caenis* decreases	Hall *et al.* 1970
18 *Chaoborus*	experimentally	*Ceriodaphnia*, *Bosmina*, *Cyclops*, *Diaptomus*, (last two are cyclopoda); under strong predation pressure, *Daphnia*	At high densities *Chaoborus* feeds on rotifers and *Daphnia*, at low densities, smaller species (*Ceriodaphnia*, *Bosmina*, *Diaptomus*) can persist, but the rotifer predator *Asplancha* (also a rotifer) is lost	Lynch, 1979†
19 *Lepomis* (fish)	experimentally	four species of *Daphnia*, *Chaoborus*, *Cariodaphnia*, *Cyclops*, *Bosmina*, *Diaptomus*	With predator, smaller species (*Bosmina*, *Cyclops*, rotifers) predominate; the rotifers support *Asplancha*. Predator removal results in larger species (*Daphnia*, *Ceriodaphnia*, the predator, *Chaoborus*). Rotifers are not sufficiently abundant to support *Asplancha*.	Lynch, 1979†

† Indicates paper where food web has been described explicitly.

There must, surely, be many real systems described by donor-controlled dynamics. Detritivores, saprophages, scavengers, indeed all species that feed strictly on dead organisms, are examples of considerable importance in view of their role in recycling nutrients. (For convenience, henceforth I shall call all dead organisms – plant or animal – detritus and the animals and plants that feed on detritus, detritivores).

Among species that feed on living organisms there are some special cases where donor-controlled dynamics would seem to be a better description than those of Lotka and Volterra. For example, seabirds restricted to feeding near a nesting colony and in water, at most, only a few meters deep will probably have little impact on their prey (fish). These special cases seem relatively few. The emphasis of this book is on species with living prey and not on detritivores. With this apology, the data of Table 5.4 are my justification in emphasizing models where predators affect the growth rates of their prey. But more importantly, the rejection of donor-controlled dynamics is also a rejection of one of the otherwise more plausible model systems where increased complexity gives increased stability.

The data in Table 5.4 also relate to a question raised earlier: how large a perturbation are real systems capable of withstanding? Clearly, the systems listed in the table do not withstand perturbations as large as species deletions. Yet all populations experience small perturbations, and the systems to which they belong persist, at least long enough to be observed. So it is on small perturbations that I shall concentrate in subsequent chapters, since these perturbations are of the most likely size to be involved in shaping food web characteristics.

5.3 SUMMARY

The theory developed in Chapter 4 suggests several possible approaches to the question of whether stability imposes upper or lower bounds on food web complexity. First, the chances of a model system being stable decline with increasing species number, connectance and interaction strength. If connectance and interaction strength are assumed constant, stability should decline with increasing species number. I discuss some attempts to demonstrate this and describe the numerous problems encountered. Most obvious is that the natural systems observed exist despite perturbations. It might be surprising if they were experimentally destabilized by small perturbations. Thus, the operational definitions of stability involve relative measures whose relationships to complexity are uncertain. Moreover, experiments have stressed the effects of species number, alone, on stability. The results are mixed and, combined with anecdotal evidence, lead to the conclusion that no clear relationship between species number and stability has been demonstrated repeatedly.

Of course, such a relationship would only be expected if species number and connectance were independent. The second prediction from the theory is that if interaction strength is independent of species number and connectance then all stable systems should be bounded by a hyperbolic function, Cn = constant. This can be easily demonstrated. Unfortunately, a much simpler explanation exists that does not invoke dynamical constraints. If species interact with a number of predators and prey which is independent of the number of species in the system, there will also be a nearly hyberbolic relationship between connectance and species number. In short, the observed relationship between connectance and species number is consistent with two models, only one of which invokes dynamical constraints.

The theory suggests a final possibility. If donor-controlled dynamics are better descriptions of the real world than Lotka–Volterra, then increased complexity would be expected to give increased stability and vice versa. For all except the simplest models, predator removals lead to further species losses with Lotka–Volterra dynamics, but no losses (or even changes in density) with donor-controlled dynamics. Nearly all studies of predator removals show either species losses or marked changes in species densities. Although there are some cases where donor-controlled dynamics would seem most likely, they are not in the majority, except perhaps for detritus–detritivore systems.

In short, I have not demonstrated that dynamics place either lower or upper bounds on food web complexity. My only conclusion is that the donor-controlled model system – the only one in which a positive correlation would be expected between complexity and stability – can be rejected for many natural systems. In Chapter 7 I shall argue that dynamics *do* place upper limits on the complexity of food webs and I shall present evidence to support that contention. In that chapter, however, I shall obtain my conclusion by examining more specific food web structures than the abstraction of 'complexity' discussed here.

Finally, recall the arguments at the beginning of Chapter 4. Whether stability places upper or lower bounds on food web complexity is only one of several questions raised by the complexity–stability problem. A second question is whether one system is more or less stable than another, but it demands a definition of relative stability. The question addressed by this chapter is whether or not stability restricts an attribute of food web structure. Another question is whether or not food web structures affect, in turn, such ecosystem functions as relative stability. It is in these general terms that McNaughton's results are interesting.

As theory suggests, McNaughton's studies showed that changes to species abundances were greater when there were more species in the system. Paradoxically, McNaughton (1977) argued that there was a positive relationship between stability and complexity. His concept of stability was a functional one: how a system changes its plant productivity in the face of changing nutrients or grazing pressure. The systems showed less change when they had more species. These results demonstrate that different definitions of complexity and stability will lead to different conclusions about their relationship. But these results also show that an ecosystem function (change in productivity) depends on an aspect of food web structure (the number of species). This raises the topic of how each of the catalogue of food web structures discussed in this book affects each of a range of ecosystem functions. This topic cannot be treated until the catalogue is complete, hence its postponement to Chapter 10. There, *inter alia*, I will discuss the relationships between functional aspects of the complexity–stability problem.

6 The length of food chains

So, Naturalists observe, a flea
Hath smaller fleas that on him prey;
And these have smaller fleas to bite 'em.
And so proceed *ad infinitum*.

(Swift, 1772)

6.1 INTRODUCTION

The energy which plants capture from the sun during photosynthesis may end up in the tissues of a hawk via a bird the hawk has eaten, the insects eaten by the bird, and the plant on which the insects fed. The plant–insect–bird–hawk system is called a *food chain*, and each stage a *trophic level*. More generally, the trophic levels are called producers (plants), herbivores or primary consumers (the insects), carnivores or secondary consumers (the bird) and top-carnivores or tertiary consumers (the hawk).

Swift's assertion is incorrect, for there are limitations to the lengths of food chains. Why? And why is the length of food chains so short, typically of three or four levels? First I shall present data to show that lengths of food chains are, indeed, shorter than would be expected by chance. Then I shall examine a number of hypotheses that have been proposed to explain this phenomenon. Some of these hypotheses make clear predictions about how food chain lengths should vary. I will evaluate these predictions as a prelude to drawing conclusions about the most likely reason for the limited length of food chains.

6.1.1 Is the number of trophic levels limited?

Food chains are never as simple as the above example suggests. That webs are more complex than the simple food chain above is a consequence of the fact that some species feed on more than one trophic level. The occurrence of such species is discussed in the next chapter. A first step, therefore, must be to establish an operational definition of the length of a food chain.

In the set of webs to be discussed in this chapter there are between one and eight species on which nothing else in the web feeds; I call these top-predators. For each I have calculated a measure which rather crudely indicates at what trophic level these species are feeding. Reference to Fig. 6.1 may help; it shows

Table 6.1 Distribution of trophic levels in ecological systems. (After Pimm, 1980b.)

n†	m	n_p	n_b	n_{int}	Tl_{mod}								C	P	*Type of system*	*Source*
					1	2	3	4	5	6	7	8				
															Terrestrial and Freshwater Systems	
10	14	5	1	25	3	3	4	3	5	–	–	–		0.140	prairie	Bird (1930)
7	8	4	3	14	4	4	4	2*	–	–	–	–		0.853	willow forest	Bird (1930)
16	21	7	2	35	3	3	4	3	3	3	3	–		C	aspen forest	Bird (1930)
23	23	6	6	48	4	4	4	2	2*	3*	–	–		C	Arctic island	Summerhayes & Elton (1923) Fig. 8.4.
3	6	4	1	9	3*	3*	3*	2*	–	–	–	–		0.144	dung and its insects	Valiela (1969)
8	11	5	2	36	4	3	3	4	4	–	–	–		0.267	freshwater stream	Minshall (1967)
6	6	1	1	14	3*	–	–	–	–	–	–	–		0.711	freshwater pond	Hurlbert, Mulla & Wilson (1972)
8	8	4	4	23	3*	3*	3*	3*	–	–	–	–	S	0.303	freshwater stream	Jones (1959)
8	11	4	1	19	3*	3*	3*	2*	–	–	–	–		0.004	spring	Tilly (1968)
10	10	3	3	15	5	4	4	–	–	–	–	–		0.696	lake fish	Zaret & Paine (1973) Fig. 6.1
7	5	1	3	7	3#	–	–	–	–	–	–	–		0.160	lake fish	Zaret & Paine (1973) Fig. 6.2
3	6	6	3	7	2*	2*	2*	2*	2*	2*	–	–		NA	tree hole (England)	Kitching (1981) Fig. 6.10
8	7	2	3	14	3*	4	–	–	–	–	–	–		NA	tree hole (Australia)	Kitching (1981) Fig. 6.10
20	22	6	4	34	4	4	4	3*	3*	3*	–	–		NA	Arctic lake	Larsson, Brittain, Lein, Lillehammer & Tangen (1978)
7	10	5	2	22	4	3	4	4	5	–	–	–		NA	Arctic lake	Gardarsson (1979)
8	13	8	3	16	2*	2*	2*	3*	2*	4*	5*	2*		HA	pitcher plant	Beaver (1979)
8	11	6	3	15	2*	3*	2*	2*	5*	2*	–	–		NA	pitcher plant	Beaver (1979)

														Marine Systems	
12	7	1	6	26	3	–	–	–	–	–	–	S	0.214	starfish	Paine (1966) Fig. 4.1(b)
6	2	1	5	7	2	–	–	–	–	–	–		–	starfish	Paine (1966) Fig. 4.1(a)
5	4	1	2	9	3	–	–	–	–	–	–		0.470	salmon and prey	Parsons & Lebrasseur (1970)
7	11	5	1	17	4	4	3	3	3	–	–	S	0.212	mudflat	Milne & Dunnet (1972)
5	9	5	1	12	3	3	3	3	3	–	–		0.100	mussell bed	Milne & Dunnet (1972)
6	9	6	3	12	3*	2*	2*	2*	2*	2*	–		NA	marine algae	Jansson (1967)
23	5	1	19	38	2	–	–	–	–	–	–		C	gastropods	Paine (1963). Simplified in Fig. 7.1(a).
12	13	2	1	31	6	6	–	–	–	–	–		NA	Antarctic sea	Knox (1970)
16	16	5	5	30	3	5	5	4	4	–	–		NA	pack-ice zone of Antarctic seas	Knox (1970)

† n is the number of species of prey; m the number of species of predators; n_p top-predators; n_b basals; n_{int} interactions; and Tl_{mod} the modal number of trophic levels. C indicates whether the web has been simplified (S) or not (blank) (see Appendix 6A). P is the proportion of random webs with n, m, n_p, n_b, and n_{int} as specified that have fewer modal trophic levels than the real web on which they are based. C indicates that the web is too complex to be analysed on existing computer facilities. NA means web was not analysed by Pimm (1980b). * Top-predators are small invertebrates which may not be at the end of the food chains; # basal species is not an autotroph.

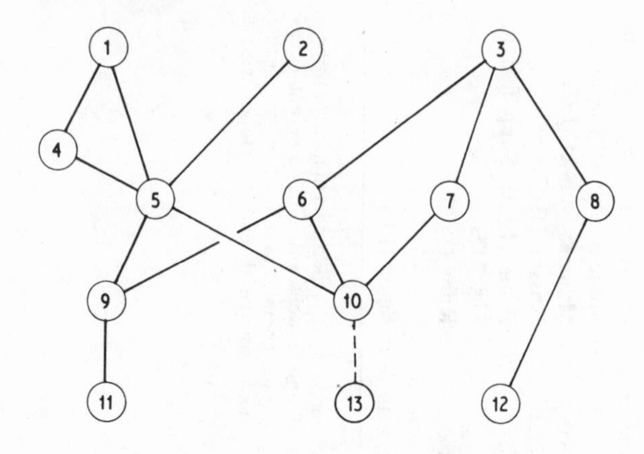

Figure 6.1 A food web of fish and their predators and prey in a tropical lake. Legend: (1) *Tarpon atlanticus* (fish), (2) *Chlidonias niger* (tern), (3) several species of herons and kingfishers, (4) *Gobiomorus dormitor* (fish), (5) *Melaniris chagresis* (fish), (6) four common species of Characinidae (fish), (7) *Gambusia nicaraguensis* (fish), (8) *Cichlasoma maculicauda* (fish), (9) zooplankton, (10) terrestrial insects, (11) nannophytoplankton, (12) filamentous green algae; (13) food of terrestrial insects. After Zaret & Paine (1973).

a web described by Zaret & Paine (1973) from a tropical lake. There are three top-predators: the fish *Tarpon atlanticus*; the tern *Chlidonias niger*; and 'several species of heron and kingfisher'.

I have placed the tern at the fourth trophic level because energy passes through three other trophic levels before it reaches the tern. The pathways used are more easily referenced by species numbers than by names. Energy reaches the tern (2) through two pathways: 11, 9, 5, 2 and 13, 10, 5, 2. For the herons and kingfishers (3) there are four pathways: 12, 8, 3; 13, 10, 7, 3; 13, 10, 6, 3 and 11, 9, 6, 3. There is a pathway of length three and three of length four. I have used a convention that a species is placed at a trophic level corresponding to the most common (modal) number of trophic levels and, in the case where there is a tie, I have placed the species at the higher (highest) of the most commonly occurring numbers of trophic levels. Consequently, the tarpon (1) is placed at trophic level five because there are pathways of lengths five, five, four and four. The choice of five is not a good one in this case. Zaret & Paine (1973) indicate that most of the energy the tarpon receives comes from species 5, which means that the pathways involving four trophic levels may be the more important ones. In most webs however, such information is lacking.

My choice of the modal number of connections is based on the supposition that, on average, this will indicate the most important path lengths. The choice of the higher value when two or more lengths occur with equal frequency is made to make more conservative the tests of whether food chains are longer than expected. These tests are described below.

The data in Table 6.1 show the modal number of trophic levels in a variety of webs. These data must be qualified in a number of ways. The species indicated at the base of the food webs are usually plants, but when plants are not at the base web I have indicated this with a # sign. These data are still useful despite their incompleteness as I shall show. The studies presented in the table were selected according to several criteria which correspond closely to those used by Cohen (1978) in his work on food webs and which are discussed in Appendix 6A.

Table 6.2 gives the frequency distribution of the number of trophic levels based on the data from Table 6.1. Commonly, systems have three or four trophic levels. What does this result mean? To say that the number of trophic levels is limited implies that the number is smaller than some *a priori* expectation. What number is to be expected? In the sense that five, six, or fifty is to be expected then the number of trophic levels is clearly limited. But there are many obvious constraints on how the species in the webs could be organized. Unless these can be met and the number of trophic levels still shown to be limited, the data cannot be taken as supporting the notion that the number of trophic levels is limited. It is possible to answer the question: how probable is the actual number of trophic levels among webs where the species' interactions are organized randomly, but subject to certain biologically reasonable constraints? I detail the methods of answering this question in the Appendix to this chapter. The method is tedious but important, since it will be used extensively in subsequent chapters. An example is sufficient for now.

Consider the second example of a lake web given by Zaret & Paine (1973) shown in Fig. 6.2(a). All the species at the base of the web are not plants (hence the # sign) and I have therefore added a putative plant as the resource for species 3. (This again is conservative, because this species, from web 6.1, is a carnivore, not a herbivore.) I ask: how should the number of trophic levels be distributed given that there are eight species in the web, seven predator–prey

Table 6.2 Distribution of trophic levels.

Number of trophic levels	*Number of cases*	*Frequencies*
2	2 (23*)	0.04 (0.25)†
3	23 (19)	0.41 (0.42)
4	24 (1)	0.43 (0.25)
5	5 (2)	0.09 (0.07)
6	2 (0)	0.04 (0.02)
Total	56 (45)	

* These are the cases where the top-predators are small invertebrates (marked with an asterisk in Table 6.1) which are probably not the end of the food chain.

† The frequencies in parentheses include those cases where small invertebrates are at the end of the food chain.

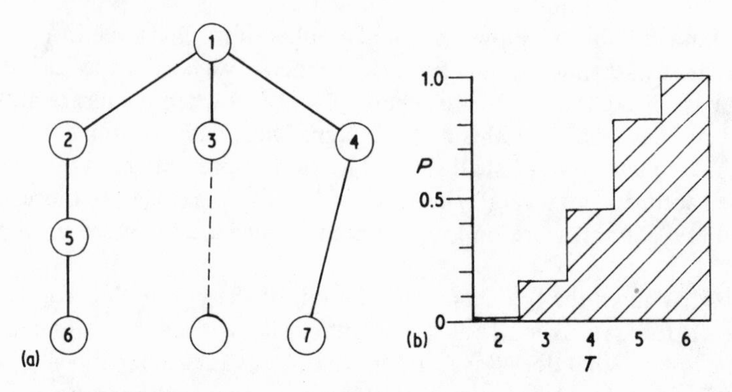

Figure 6.2 (a) Tropical lake fish and their prey. Legend: (1) *Cichla ocellaris* (fish), (2) young *Cichla*, (3) *Gobiomorus dormitor* (fish), (4) *Cichlasoma maculicauda* (fish), (5) zooplankton, (6) nannophytoplankton, (7) filamentous green algae. Unnumbered: prey of species (3). After Zaret & Paine (1973). (b) The proportion (P) of model food webs that have as many, or fewer, trophic levels (T) yet have the same number of top-predators (1), basal species (3), prey (7), predators (5), and predator–prey interactions (7) as the web in (a).

interactions, one top-predator and three species at the base of the web? There is one additional constraint which prohibits loops of the kind A eats B eats C eats A.

The distribution shown in Fig. 6.2(b) was obtained by computer. The real web has only three trophic levels yet the randomly generated models can have many more. Indeed, the proportion of randomly generated webs that had three trophic levels or less was 0.160. This process, described in more detail in the Appendix, was undertaken for many of the webs in Table 6.1. If real webs had no limitation on the number of trophic levels, the expected proportion would be 0.5. That is, for half of the time the real webs would have fewer and for the other half of the time more trophic levels than the random webs. For the webs in Table 6.1, the mean value of the proportion is 0.333, a value that is significantly less than 0.5. Real webs do have fewer trophic levels than would be expected by chance. Why should this be so?

6.2 HYPOTHESIS A: ENERGY FLOW

Elton (1927) wrote: 'the animals at the base of the food chain are relatively abundant while those at the end are relatively few in number, and there is a progressive decrease between the two extremes'. This is the 'pyramid of numbers' and it is a widespread observation though there are some notable exceptions. All these observations are reconciled by considerations of energy flow.

Lindemann (1942) was one of the first to view ecosystems in energetic terms, and his view of trophic levels is summarized in Fig. 6.3. Of the radiant energy

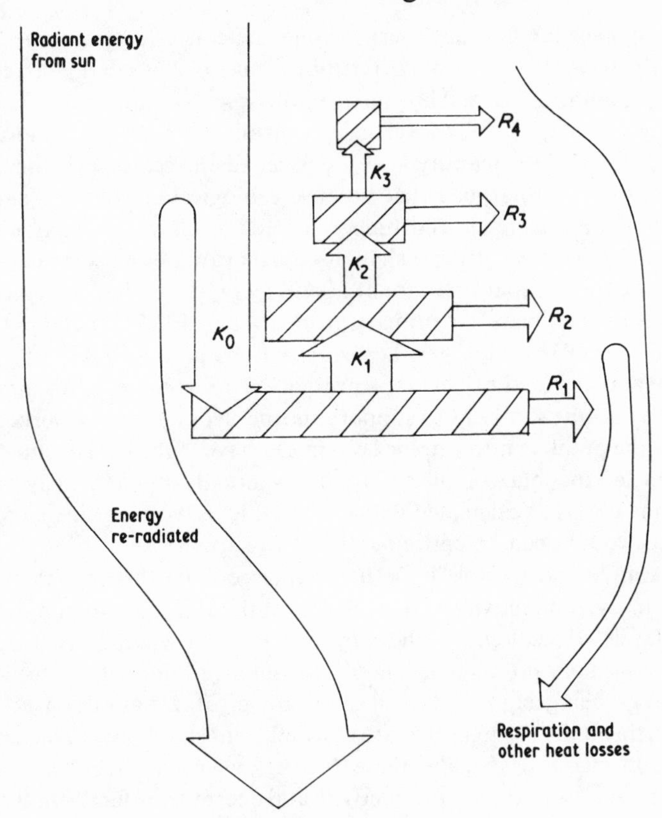

Figure 6.3 The flow of energy through an ecosystem. Of the radiant energy from the sun only a portion (K_0) is used by the plants. Of this, some is lost as heat during the night when the plants respire (R_1) and some energy is converted into plant material consumed by the herbivores (K_1). Of K_1, some is lost through respiration (R_2) and some is used by the carnivores when they feed on herbivores, and so on. After Phillipson (1966).

reaching the earth from the sun, only a tiny fraction is used by plants in photosynthesis. Some of the energy absorbed by the plants is lost when the plants respire during the night, but some of this energy, k_1, will be used by the animals which feed on the plants. Of this energy, some will be lost by the animals as they respire, but some, k_2, will be passed on to the next trophic level and so on throughout the food chain. Some energy will be lost as waste (faeces and urine) and some will be used in changing the amount of biomass at each level (the standing crop) if the biomass increases. At equilibrium, by definition, biomass will not increase. The first law of thermodynamics states that the total amount of energy remains constant; when one animal eats a plant the energy is merely converted from one form into another. But the second law of thermodynamics implies that energy cannot be converted from one form to another (from victim to consumer) without some of this energy being lost as heat – the respiration term. This raises an obvious question: what fraction of the energy entering a trophic level, k_{n-1}, is available to the species at the next trophic level, k_n? The quantity k_n/k_{n-1} is called the ecological efficiency.

Hutchinson (1959) argued that because ecological efficiency is small, the energy available to higher and higher trophic levels will quickly diminish. Suppose the efficiency is 10% (I shall discuss its variation in some detail later) and suppose further that the amount of energy produced by the plants that is available for the animals (net primary production) is 10^4 kcal per unit time per unit area. Then 10^3 kcal of new herbivore will be produced, 10^2 kcal of new carnivore and 10^1 kcal of new top-carnivore. Unfortunately, this amount of energy may not be sufficient to support another trophic level. Alternatively, a species at the putative fifth trophic level might have to feed over a much larger area in order to obtain sufficient food. Eventually the area may run out. Slobodkin (1961) has calculated that at trophic level twenty a single predatory individual would need an entire continent to support it.

Some simple and testable hypotheses emerge from this argument. As an area's primary productivity is increased it should be able to support more trophic levels. If ecological efficiency is 10%, every order of magnitude increase in primary productivity should permit an additional trophic level, all other things being equal. Also, if there are great variations in ecological efficiency, then those systems that are more efficient should be characterized by longer food chains. To make these predictions more precise I shall first consider patterns of primary productivity and energy transfer. With these data available it will be possible to state more precisely some testable hypotheses.

6.2.1 Patterns of primary productivity

Primary productivity is known to vary widely. For marine planktonic systems, Koblenz–Mishke, Volkovinsky & Kabanova (1970) report values from 0·1 to 10^2 mg C (carbon) m^{-3} day^{-1}. In benthic habitats Bunt (1975) cites 16 studies with values ranging from 10 to 2×10^4 mg C m^{-2} day^{-1}. Of these, seaweed-dominated habitats are among the most productive with values in the range of 1.65×10^3 to 2.0×10^4 mg C m^{-2} day^{-1}. For 12 studies of freshwater lakes, Likens (1975) cites high latitude lakes with productivities as low as 14 mg C m^{-2} day^{-1} (Lake Vanda, in Antarctica, covered by 4 m of ice), temperate latitude lakes with intermediate values and some tropical lakes with values as high as 1.75×10^4 mg C m^{-2} day^{-1}. Leith (1975) summarizes data for terrestrial studies and original sources are not given. None the less the range of values is large. Tropical rain forests and some marshes and swamps may reach 4.6×10^3 mg C m^{-2} day^{-1}, while other woodlands and grasslands span the range 230 to 2.3×10^3 mg C m^{-2} day^{-1} and deserts from 10 to 280 mg m^{-2} day^{-1}. Arctic tundras range from 115 to 460 mg C m^{-2} day^{-1}. The data were given as annual productivities and I have divided by 365 to obtain daily productivity, which means that some daily values will exceed these values when the growing season is short. But if it is supposed that the productivity is concentrated into only one-third of the year (and these daily productivities are therefore multiplied by three) then the large range of productivities still persists. Moreover, these ranges are probably under-estimated. In more detailed studies that give the exact sources of data, comparable habitats are given larger ranges. For example, Murphy (1975) shows tropical grasslands ranging from 170 to 4400 mg C m^{-2} day^{-1}. Similarly, tundra ecosystems are found with productivites outside Leith's ranges. For example, Bliss (1977) describes three systems with productivities as low as 8 mg C m^{-2} day^{-1}.

To summarize, the productivities of marine, freshwater and terrestrial ecosystems span four orders of magnitude, approximately 10^1 to 10^4 mg C m^{-2} day^{-1}, as do marine planktonic systems, which range from 10^{-1} to 10^2 mg C m^{-3} day^{-1}. Certainly, there is the potential to find considerable variation in food chain lengths if they are affected by primary productivity.

6.2.2 Trophic efficiencies

Ecological efficiencies are measured less often than the ratio of production (P) to assimilation (A = production plus respiration, R). Because waste (such as faeces and urine) is ignored, the P/A ratio overestimates ecological efficiency and does so by an amount which will differ depending on the proportion of energy consumed that is assimilated (assimilation efficiency). The P/A ratio must be an approximate measure of ecological efficiency when respiration is the major use of energy, as it is for many species. Simply, variation in assimilation efficiency is small enough relative to that of P/A so that the variations in ecological efficiency will closely mimic those in the P/A ratio.

The P/A ratio was studied by McNeill & Lawton (1970) and by Humphries (1979). The general results are the same in both studies, in that there are large differences in P/A among different taxonomic groups. Endotherms (birds and mammals), with their additional requirements of keeping warm in cold weather, are the least efficient, with P/A being usually less than 3%. The only vertebrate ectotherms measured, fish, have efficiencies of about 10%. Non-insect invertebrates are more efficient, with efficiencies in the range of 21–36%, and insects are the most efficient, with mean values of 39% for herbivores and 56% for carnivores. These efficiencies do not change as assimilation increases, although assimilation does vary over nearly five orders of magnitude.

6.2.3 Three predictions

With these data available, three predictions can be made:

(1) Sites with high primary productivities should have more trophic levels than those with low productivities. This is conditional on the sites having a similar composition of trophic groups (endotherms versus ectotherms, etc.).

(2) Using data on the standing crop of a particular trophic level and the primary productivity of the systems, it should be possible to predict a threshold value below which primary productivity will be insufficient to support that trophic level.

(3) Suppose that at a particular trophic level there are (i) insects (ii) other invertebrates (iii) ectothermic vertebrates or (iv) endotherms. Then more trophic levels should be able to follow (i) than (ii) than (iii) than (iv) because this order reflects the availability of energy to subsequent trophic levels.

(a) Prediction 1: effects of primary productivity

Grasslands and tundras have some obvious features in common: they are terrestrial, often treeless ecosystems. Conveniently, they have been studied extensively as part of the International Biological Programme (IBP). Their faunas are relatively small and reasonably well known. Two studies of the tundra are those of Wielgolaski (1975; Fennoscandian tundras) and Bliss (1977; Canadian tundras). The first author describes three sites, one of which, Kevo in Finland, has the lowest annual primary productivity, averaging 440 mg C m^{-2} day^{-1}. None the less, three trophic levels are common. There are several species of endothermic predators which feed on endothermic herbivores (this is the least efficient pathway for energy flow). There are also Merlins (*Falco columbacius*)–small falcons–which probably feed on species of insectivorous birds, placing them at the fourth trophic level. The tundra studied by Bliss has two principal habitats: raised beaches (96 mg C m^{-2} day^{-1}) and hummocky sedge meadows (220 mg C m^{-2} day^{-1}). Found there are mammals: weasels, foxes (which feed on birds, which are partly insectivorous), wolves and Polar Bears (*Thalarctos maritimus*). And birds: passerines, as well as Snowy Owls (*Nyctea scandiaca*) (which eat lemmings, ducks and some skuas) and Peregrine Falcons (*Falco peregrinus*) (which feed on passerine birds, sandpipers and skuas). Skuas are, in turn, predatory. The systems clearly support three trophic levels and there is a suggestion of a fourth.

The North American grasslands studied by French (1979) have primary productivities of 100 to 640 mg C m^{-2} day^{-1}. There is a measurable third trophic level at each of the sites described (discussed in more detail below) and my observations (Raitt & Pimm, 1976; unpublished) suggest a fourth may be present. The sites have one or more of: Prairie Falcons (*Falco mexicanus*) (which feed, in the summer, on insectivorous birds), Hen Harriers (*Circus cyaneus*) (also a bird-predator) and Swainson's Hawks (*Buteo swainsoni*) (which feed, in part, on snakes and insectivorous lizards), and snakes which consume the nestlings of the insectivorous birds.

One of the most productive grasslands is in West Africa and has a primary productivity of about 1900 mg C m^{-2} day^{-1} (Lamotte, 1975). There is a well defined third trophic level of insectivorous birds, amphibians, lizards and one rodent. There is also a fourth trophic level of snakes which feed on the lizards and amphibians.

These data show a range of habitats with productivities ranging over an order of magnitude, yet their trophic structures are almost identical: all have a definite third trophic level with some species at the fourth trophic level. In all but one, animals at the fourth trophic level are endothermic and feed on other endotherms; the exception – a vertebrate ectotherm feeding on other vertebrate ectotherms – occurs at the most productive site, not the least. The most productive IBP site I could find is a Czech fish pond, its margins having a productivity of at least 4000 mg C m^{-2} day^{-1} (Dykyjova & Kvet, 1978). This is close to the upper limit for terrestrial systems, yet there is no indication of more than four trophic levels. Indeed, four trophic levels seem no more common than at the barren tundra sites already described.

A similar pattern emerges from freshwater studies. Lake Myvatn in Iceland (Johanssen, 1979) has a primary productivity of 25 mg C m^{-2} day^{-1} which comes largely from benthic plants. Yet, the study includes a food web with a clear fourth trophic level containing fish-eating ducks. Similarly, Lake Øvre Heimdalsvatu in Norway has a productivity (plus organic input from streams and rivers) of no more than 120 mg C m^{-2} day^{-1} and also has a clear fourth trophic level (Larsson, Brittain, Lein, Lillehammer & Tangen, 1978). The eutrophic Lake Suwa in Japan (Mori & Yamamoto, 1975) has phytoplankton which alone produces 1180 mg C m^{-2} day^{-1}, yet no indication of a fourth trophic level is given. Finally, there is tropical Lake George in Uganda with a primary productivity of 1240 mg C m^{-2} day^{-1}. Its food web is described explicitly and includes species at a fourth, but not a fifth, trophic level, (Burgis, Darlington, Dunn, Ganf, Gwahaba & McGowan, 1973, Moriarty, Darlington, Dunn, Moriarty & Tevlin, 1973.)

Finally, because man is at the end of a food chain and because our food supply is well documented, it is interesting to ask: at what trophic level does man feed? A casual visit to the market shows that our terrestrial food consists almost entirely of plants and herbivores. Rather more interesting are the marine fish we exploit. Ryther (1969) has grouped marine ecosystems by their productivities: open oceans ($\sim$ 50 g C m^{-2} yr^{-1}), coastal areas ($\sim$ 100 g C m^{-2} yr^{-1}) and upwellings ($\sim$ 300 g C m^{-2} yr^{-1}). Man feeds at trophic levels five, three and two for these three systems respectively. We put tuna (a top-predator from the open ocean) in our sandwiches but anchovies (herbivores from upwellings) on our pizzas. Simply, man's food chain is not longer in the more productive systems, but actually appears to be shorter.

In summary, there is no evidence that food chains are longer in more productive habitats. Nor is there any evidence that less productive habitats show compensations by being characterized by energetically more efficient groups such as ectotherms.

(b) Prediction 2: the existence of a threshold primary productivity

Several of the grassland, tundra and lake studies quoted above provide a quantitative estimate of both the standing crop of the third trophic level and the primary productivity. Their relationship is shown in Fig. 6.4, in which I have retained the original units of grams dry weight per square metre per year. (One gram C is equivalent to approximately 2.4 grams dry weight.) At between 40 and 50 g dry weight m^{-2} yr^{-1} it can be predicted that habitats would be unable to support a third trophic level. For comparison, this is an average of about 50 mg C m^{-2} day^{-1}. Two points are below this threshold. One is the Antarctic Lake Vanga, which from descriptions (Goldman, Mason & Hobbie, 1967) apparently does lack a third trophic level, and the other is one of the tundra systems described by Bliss (1977), which has a third trophic level. There is also anecdotal evidence of other habitats with very low productivities supporting a third trophic level. My favourite example is Thesiger's (1959) description of the plant–camel's milk–man food chain in one of the earth's most desolate regions – the Empty Quarter of Saudi Arabia. None the less, the data strongly suggest that below a certain level, support of a third trophic level must be difficult. In the extreme, low productivities must limit the length of food chains.

(c) Prediction 3: trophic group and food chain length

From the data in Table 6.1 I have extracted each food chain that appears complete. All but four have invertebrates at their second trophic level. Species at the third trophic levels can be classified into one of three groups: endotherms, vertebrate ectotherms and invertebrate ectotherms. For each group I have calculated the mean number of trophic levels in excess of the third trophic level. The prediction is that food chain lengths should be shorter following endotherms than vertebrate ectotherms than invertebrate ectotherms. The mean lengths are 0.625 ($n = 8$), 0.583 ($n = 12$) and 0.583 ($n = 12$) for the three groups respectively. Hence, there is no tendency for food chains to be longer when built above energetically efficient groups, like invertebrates, than when built above energetically profligate groups like endotherms.

6.3 HYPOTHESIS B: SIZE AND OTHER DESIGN CONSTRAINTS

The size of an animal has the potential to limit the length of food chains in a number of inter-related ways. Generally, a predator will be larger than its prey (Hutchinson (1959) suggested twice as large), and conversely a parasite must usually be smaller than the prey on (or in) which it feeds. There are exceptions,

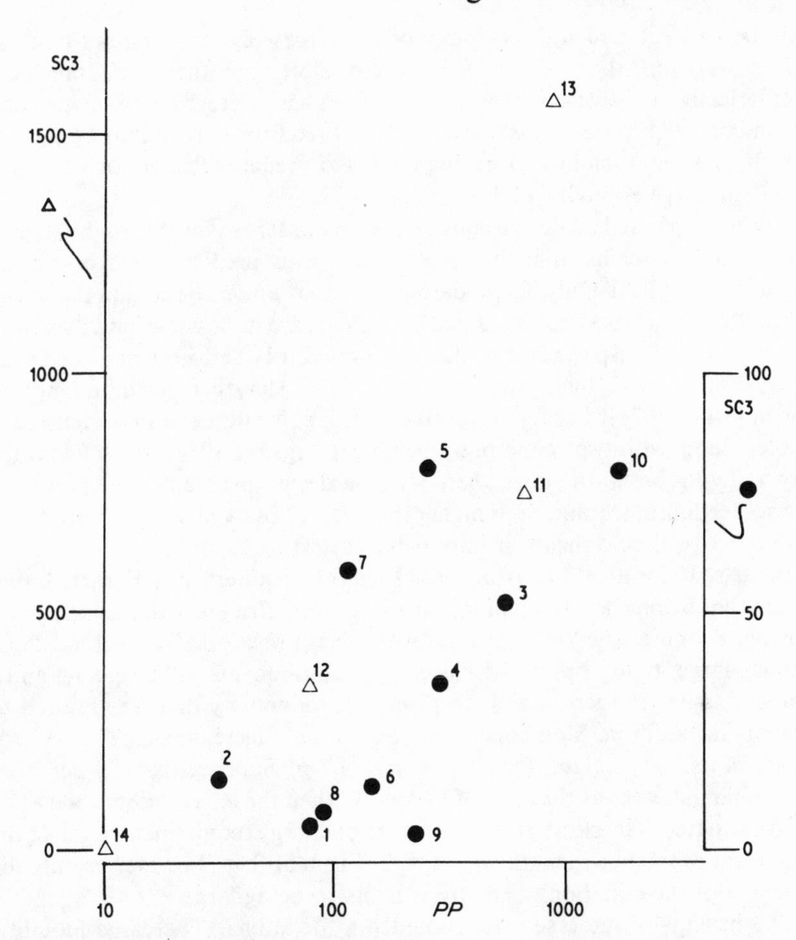

Figure 6.4 Standing crop (in mg dry weight per m^2) of secondary consumers (SC3) against primary productivity, PP (g dry weight per m^2 per year) for terrestrial (circles, scale at right) and freshwater (triangles, scale at left) ecosystems. Legend: (1) tundra, sedge moss, (2) tundra, raised beach, (3) Osage, (4) Cottonwood, (5) Pantex, (6) Pawnee, (7) Jornada, (8) ALE, (9) Bridger, (10) West Africa, (11) Lake Myvatn, (12) Øvre Heimdalsvatu, (13) Lake George, (14) Lake Vanda. Sources: (1) and (2), Bliss (1977); (3) through (9) are International Biological Programme sites in the USA discussed by French (1979); (10) Lamotte (1975); (11) Johanssen (1979); (12) Larsson, Brittain, Lein, Lillehammer & Tangen (1978); (13) Burgis, Darlington, Dunn, Ganf, Gwahaba & McGowan (1973); (14) Likens, (1975).

however, such as lions and tigers feeding on larger mammalian herbivores. Predators must be 'fiercer' than their prey, where ferocity involves a complex of physical attributes (including size, strength and speed) as well as behavioural ones. All these factors impose 'design restrictions' on the species. Birds can only be so large and still fly. Many of the largest flying predators (such as eagles) rely on thermals to keep them aloft, for their size precludes energetically expensive, flapping flight. That Peregrine Falcons and Gyrfalcons (*Falco rusticolus*) have no avian predators may well be because of the physical impossibility of having a winged predator fast enough to catch them and large enough to kill them.

Arguments based on design constraints encounter two problems. First, they are *ad hoc*: it seems unlikely that one can ever predict which beasts are inevitably mythical. Only if a predator does exist can one be certain that it can be 'built'. Second, best guesses at which animals are impossible have a habit of being wrong. I suspected once that hummingbirds had no avian predators because of their fast, highly manoeuvrable flight. Alas, there is a hummingbird falcon (Stiles, 1978)! The fossil record contains some outrageous designs, such as pterosaurs with twelve-metre wingspans (Langston, 1981) which dwarf the largest living birds. In short, there may be design problems that prevent a species feeding at trophic levels higher than those observed, but this hypothesis is one that will be difficult or impossible to test.

Increased size also means increased energy requirements. Earlier, I suggested that trophic levels might increase linearly with a logarithmic increase in primary productivity. Yet if the putative species to be added to the food chain is much larger than its prey, then its energy requirements will be greater and a proportionately larger increase in primary productivity may be required to support its addition. Metabolic costs of an animal increase as the two-thirds power of its body weight (for birds: Pimm, 1976). Suppose that the weight of an animal increases as the cube of its length; then the energy costs scale as its length squared. The slope of the logarithm of energy requirements against the logarithm of the body length would be 2 (Fig. 6.5). This, however, is only the cost side of the equation; there are benefits to being large.

The benefits of increased size come from the animal's increased mobility. Larger animals roam over larger areas than smaller ones and have, as a consequence, at least the potential for increased supplies of energy. Suppose an animal's energy supply is directly proportional to its home range. Hutchinson & MacArthur (1959) and Southwood (1978) have shown that an animal's home range increases with its size. They used the diameter of the home range. (Clearly, the area increases proportionally to the square of this, or twice as fast on a logarithmic scale.) Interestingly, both sets of authors show that the diameter of the home range plotted against the length of the animal (both expressed in logarithm) has a slope of 1 (or perhaps slightly higher). Thus, the area exploited by an animal increases approximately as the square of its length, that is, the same relationship as for energy costs (Fig. 6.5). In short, the benefits of increased size seem to increase exactly as fast as the costs of increased size; there is apparently no energetic disadvantage to being large.

This analysis is superficial as many factors other than size contribute to the size of an animal's home range. In addition, Southwood (1978) shows that

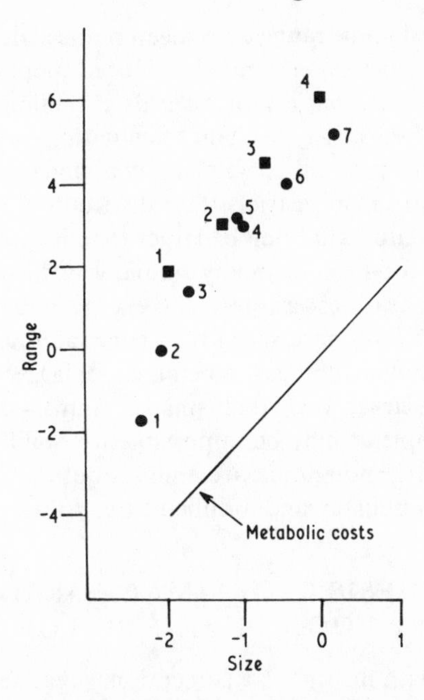

Figure 6.5 Consequences of size to the energy budget of animals. Symbols: $\log_{10}$ of an animal's home range (R) in m^2 against $\log_{10}$ of an animal's length in m. Legend: squares (carnivores): (1) beetle, (2) shrew, (3) weasel, (4) fox; circles (herbivores): (1) fly, (2) grass bug, (3) grasshopper, (4) vole, (5) mouse, (6) rabbit, (7) deer. After Southwood (1978). Line: metabolic costs (on an arbitrary scale) against an animal's length. There is no significance to the line being below the symbols because the line's scale is arbitrary (I lack the information to convert the animal's home range to the energy it can obtain from that area). However, the ratio of an animal's costs to benefits (probably proportional to home range) does not increase as its size increases.

increased size is often accompanied by a diversification of diet, suggesting that larger species have potentially more food available to them. But a greater variety of food supplies may be accompanied by a reduction in the efficiency with which the animal exploits those resources. How these various factors modify the ratio of cost to benefit with changing size is clearly impossible to guess. It is interesting, therefore, that Damuth (1981) has shown that the energy per unit area used by populations of mammalian herbivores is independent of body size over some five orders of magnitude of body size. His results are derived from density estimates which decline with body size at a rate that exactly balances the increasing energy requirements of individuals of larger size. This independent approach confirms that large size *per se* does not make a species impossible energetically. The results reject the notion that a species' share of the energy pie might decrease if the individuals evolved towards a larger body size.

Of course, increased home range does mean reduced density and, when the area suitable for a species is limited, reduced population sizes. Small populations suffer from a variety of hazards including genetic ones from inbreeding and undue sensitivity to chance fluctuations in numbers. Species at the top of food chains may still be sufficiently abundant to be beyond these problems, at least, until man restricts their distribution (Newton, 1979).

In summary, a predator will often be larger than its prey, but size does not restrict the length of food chains in any simple way. Size affects at least two important biological features: an animal's energy budget and its physical ability to perform the tasks necessary for its survival. Energetically, there are both disadvantages and advantages in being large: larger animals need more energy but feed over larger areas than smaller animals. Size imposes design constraints, for example, it may be impossible to 'build' a flying Gyrfalcon predator. Yet, there are enough bizzare animals both living and fossilized to make this hypothesis unlikely and a difficult one to test.

6.4 HYPOTHESIS C: OPTIMAL FORAGING; WHY ARE FOOD CHAINS SO LONG?

Hutchinson (1959), with his singular perception, suggested that the question addressed so far might be the wrong one. Perhaps it should be: why are food chains so *long*? There is more energy available at the $(n-1)$th trophic level than at the nth level for reasons already discussed. Indeed, if the ecological efficiency is 10%, there will be an order of magnitude more energy at the lower trophic level. Hastings & Conrad (1979) have extended this argument. They considered a predator's strategy to be optimal if it exploited the thermodynamically richest accessible environment. Consider a fourth level species which feeds primarily on carnivores. Its nutritional needs could be satisfied at a lower trophic level by herbivores because carnivores and herbivores are likely to be nutritionally interchangeable. In the absence of additional non-nutritive constraints, members of this species could always compete with their former carnivore prey for an energetically richer resource, namely, the herbivores. The option of switching to a lower trophic level is not available to the first level carnivore because of the nutritional differences between plants and herbivores. Hastings & Conrad concluded 'that in the absence of non-nutritive constraints grazing food chains should evolve towards an evolutionary stable length of three.'

This argument is pleasingly simple and there are some possible examples. The phytoplankton–zooplankton–baleen whale food chain of the Antarctic convergence is one that comes to mind. Unfortunately, the argument is not consistent with the high frequency of four trophic levels in the real world (Table 6.2). And, in contrast to Hastings & Conrad's argument, there are good reasons why species should feed on the highest trophic level available to them. The existing carnivore might be very efficient at exploiting its herbivorous prey; so efficient, in fact, that the new species might gain more from feeding on the carnivore than competing with it for the herbivorous prey. Hastings & Conrad's model is true only if the existing carnivore is a truly inept competitor. Moreover, once a top-carnivore has established itself, the strategy for the next invading species is even less certain. If it tries to exploit the herbivores, it not only must face competition from the resident carnivore but also the threat of predation from the top-carnivore. If the new species enters at trophic level four, it has a competitor in the resident top-carnivore. Yet, if it enters as the top-carnivore's predator, it has none of these problems, though its potential food supply is small. I shall show in Chapter 9 that the lower in the food chain a species feeds, the more predators it suffers and the fewer prey it exploits. There may be good reasons to feed low in the food chain, but there are also reasons to feed high in the food chain. And so long as there are evolutionary tendencies both to lengthen and shorten food chains, then an evolutionary equilibrium will result. This will still result in food chains shorter than would be expected by chance, and how they will vary with primary productivity is unclear.

6.5 HYPOTHESIS D: DYNAMICAL CONSTRAINTS

The requirement of dynamical stability imposes restrictions on the length of food chains. There are three arguments. The first is that as food chain lengths increase it becomes less likely that model systems will have feasible equilibrium densities. This will be seen to be a repetition of the energy flow argument. Second, for feasible systems, long food chains are unstable in stochastic environments. Third, systems modelled by difference equations are less likely to be stable the more trophic levels they have.

6.5.1 Feasibility

From the Lotka–Volterra models for one, two and three species in linear chains, the conditions can be calculated that ensure that all the one-, two- or three-species equilibrium densities are feasible (Pimm, 1979a). Choosing the signs of the coefficients appropriately so that all the parameters are positive yields

$$
\begin{aligned}
&b_1/a_{11} > 0\\
&b_1/a_{11} > b_2/a_{21}\\
&b_1/a_{11} > (b_2/a_{21}) + (a_{12}b_3)/(a_{32}a_{11}a_{21}). \qquad (6.1)
\end{aligned}
$$

It can be seen from these expressions that, for a fixed b_1/a_{11}, it becomes increasingly difficult to satisfy the conditions as trophic levels are added. This argument is the energy flow argument in another guise. The term b_1 is the rate of production of new individuals of X_1; a constant (the number of kcal per individual) converts it to an energy flow. The term a_{11} similarly can be viewed as the energy wasted by intraspecific competition and therefore denied to subsequent trophic levels. The ratio of these terms is related to the amount of energy available to subsequent trophic levels. From Expressions (6.1), the larger this ratio, then the more trophic levels can be supported.

6.5.2 Food chain lengths in stochastic environments

Assuming that energy flow is sufficient to satisfy the constraints discussed above, there is still a limitation imposed by population dynamics. Consider all the n-species models with one species per trophic level and with the species at the base of the web limited by some resource (Fig. 6.6). These models are qualitatively stable, irrespective of the value of n, and are also globally stable, (Harrison, 1980). None the less, long food chains are dynamically fragile.

Pimm & Lawton (1977) used Lotka–Volterra models to construct food chains of varying length and judged their fragility by how long the populations took to return to equilibrium following a disturbance. Recall that long return times (eigenvalues negative, but close to zero) imply an increased probability of

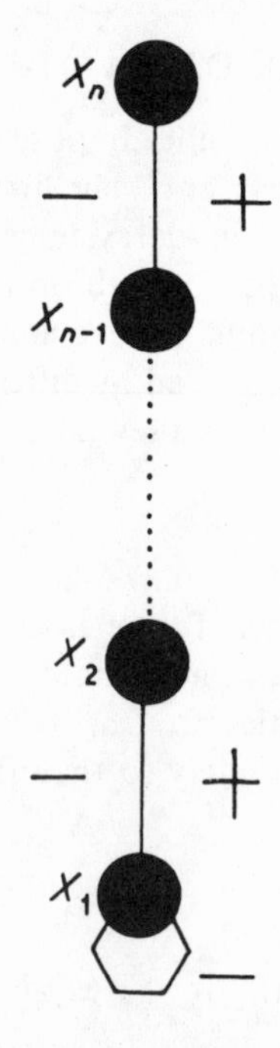

Figure 6.6 A general model of n species, with one species per trophic level. Such models are locally and globally stable. For simplicity, predator–prey interactions are represented with a single line but with two signs (+, −) to indicate that both predators affect prey (−) and vice versa (+). Only species at the base of the web affect their own growth rates (−).

extinction for some of the species in a world where there are stochastic changes in species densities. The eigenvalues of matrices whose elements were chosen randomly subject to the constraints of sign and magnitude were investigated in this study. The models investigated are shown in Fig. 6.7. All models had four species, but arranged into two, three and four trophic levels. Recall that the elements of the Jacobian matrices, from which the eigenvalues are calculated, are $a_{ij}X_i^*$. For predator–prey interactions, a_{ij} is the rate of new predators produced per prey eaten or the number of prey killed by the addition of a predator; the latter will usually be much larger than the former. The X_i^* will also differ, in that predators are likely to be fewer in number than their prey at equilibrium. To model these asymmetries, the effects of the predators on the prey were chosen randomly and uniformly over the interval (−10, 0) and the effects of the prey on the predators similarly over the interval (0, +0.1).

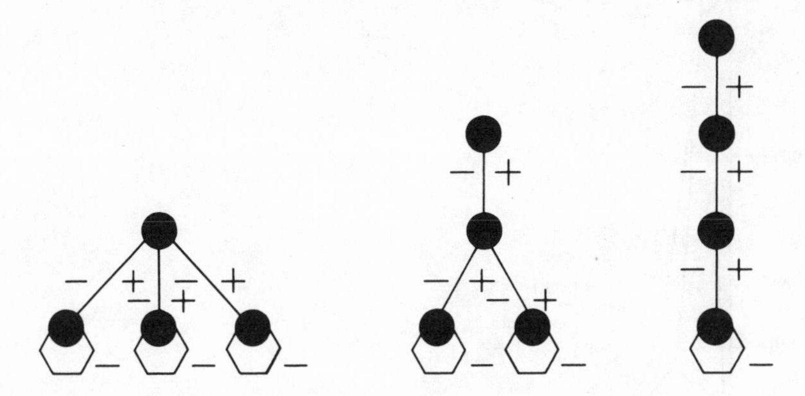

Figure 6.7 Digraphs of three four-species systems with two, three and four trophic levels. For sign conventions see Fig. 6.6.

For each of the three models shown in Fig. 6.7, 2000 matrices were created, and Fig. 6.8 shows the distribution of their return times. As the number of trophic levels increases, so do the return times. (Recall that return times, defined by Equation 2.42 are the times taken for a perturbation to decay to 1/e of its initial value. The units are those chosen to express the rate of change of the population in the model). For two trophic levels, nearly 50 % of the models had return times less than or equal to 5, falling to 5 % for three trophic levels and to less than 1 % for four. Also indicated in the figure is the percentage of models where return times exceed 150. With two trophic levels this is very small (0.1 %), but equal to 9 % and 34 % respectively with three and four trophic levels.

An alternative approach is to have only one species per trophic level (as in the case of the four-species, four trophic level model). For a two-species model with two trophic levels, the distribution of the return times can be analysed

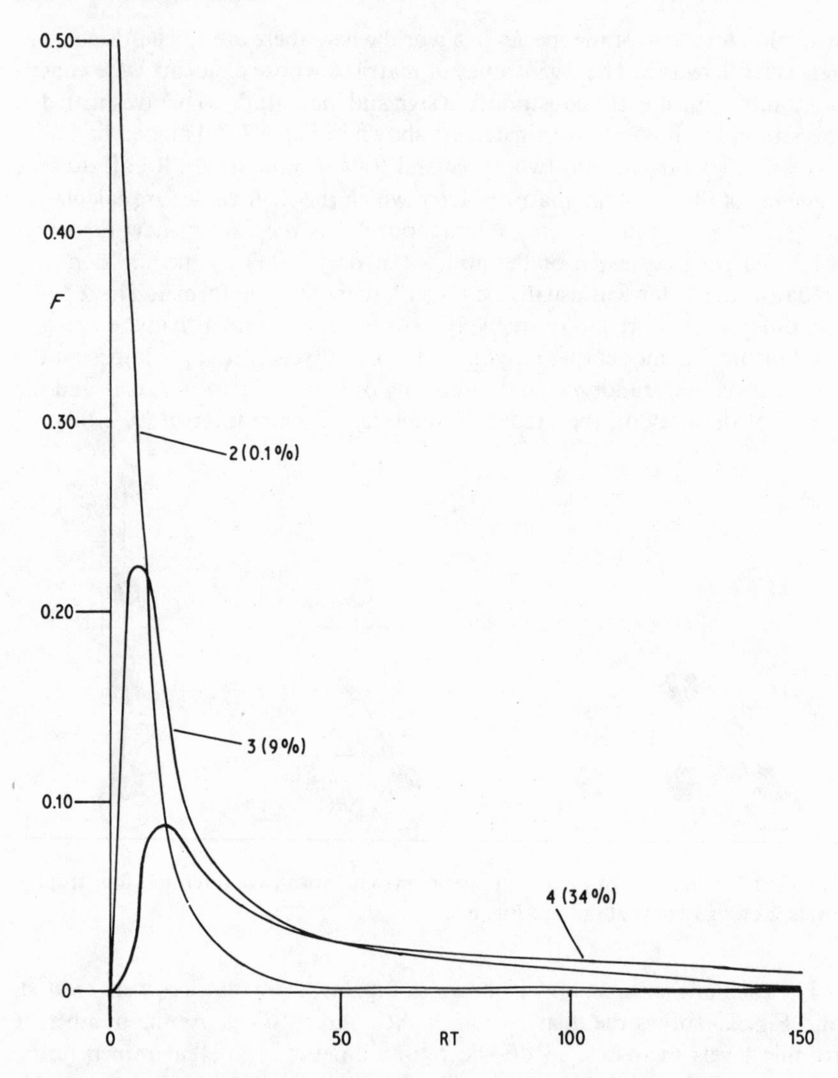

Figure 6.8 Frequency (F) of return times (RT) for the two, three, and four trophic level models shown in Fig. 6.7. Percentages are for models with return times in excess of 150. After Pimm & Lawton (1977).

analytically. Here about 70% of the models had return times of less than 5. Hence, models with different numbers of species and the same number of species per level, as well as models with the same number of species, lead to the same conclusions: as the number of trophic levels increases so do the average return times. Because systems with long return times are unlikely to persist in a stochastic world this suggests that only systems with a few trophic levels will be found commonly in nature.

This result has been derived analytically by Vincent & Anderson (1979) for Lotka–Volterra models in general. They also investigated the property of vulnerability discussed in Chapter 3. Longer food chains are not, inevitably, more vulnerable to perturbations than shorter ones, but often longer return times mean increased vulnerability to the most malevolent perturbations a system will encounter.

6.5.3 Food chain lengths in finite difference systems

For differential equations it is possible to produce model food chains of any length yet which have entirely negative eigenvalues. These, in a totally predictable environment (were it to occur!), would be stable. The conditions for stability in finite difference equations, however, are more restrictive, and for these models long food chains are not likely to be stable. One study, by Beddington & Hammond (1977), has investigated the effect of adding a third trophic level to a two trophic level system. The models deal with an insect host (X_1), its parasitoid (X_2) and a hyper-parasitoid (X_3) which attacks the parasitoid. The model used by the authors has the general formulation

$$X_{1,t+1} = X_{1,t} f_1(X_{1,t}) f_2(X_{2,t})$$
$$X_{2,t+1} = X_{2,t}[1 - f_2(X_{2,t})] f_3(X_3, t)$$
$$X_{3,t+1} = X_{3,t}[1 - f_2(X_{2,t})][1 - f_3(X_{3,t})], \quad (6.2)$$

where X_i are the densities of the three species at times t and $t+1$, the function $f_1(X_{1,t})$ describes the host's rate of increase in terms of its own density, $f_2(X_{2,t})$ describes the proportion of hosts surviving the attentions of the parasitoid and $f_3(X_{3,t})$ describes the proportion of the parasitoids that survive the attack of the hyper-parasitoid. This model assumes a sequence of events in which the hosts parasitized by X_2 then become susceptible to parasitism by X_3. The precise forms for the functions are

$$f_1 = \exp[b_1(1 - a_{11} X_1)]. \quad (6.3)$$

This is one of a number of difference equation analogues to the single-species differential equation discussed earlier (e.g. Equations (2.15), (3.8), (3.9)). In the absence of predation, if $X_t = 1/a_{11}$ then $X_{t+1} = X_t$ and the rate of approach to equilibrium is given by b_1.

$$f_2 = \exp(-a_{12} X_{2,t}) \quad (6.4)$$
$$f_3 = \exp(-a_{23} X_{3,t}). \quad (6.5)$$

These forms were described in Chapter 2 and are derived by assuming that the parasitoids search randomly for hosts, with a_{ij} $(i \neq j)$ representing the so-called 'area of search', that is, a measure of the proportion of the hosts parasitized at a given parasitoid density.

Unlike the differential equation models, not all parameter values consistent with a feasible equilibrium give stability. Figure 6.9 shows the limited subset of

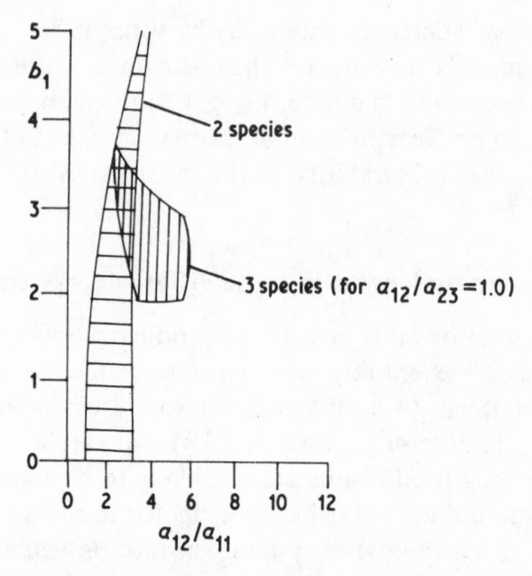

Figure 6.9 Combinations of parameter values (see Equations (6.3)–(6.5)) that give stability for two trophic level systems (horizontal shading) and three trophic level (vertical shading) insect host–parasitoid systems. After Beddington & Hammond (1979).

values that lead to stability for the two-species system, without the hyper-parasitoid. The figure also shows the region of parameter combinations where a three trophic level system will be stable. The example is but one of an infinite number of possible combinations for the ratio a_{12}/a_{23}. It is difficult to draw the entire three-dimensional diagram clearly. The figure, however, is quite typical in one important respect: the set of parameters consistent with stability for a three trophic level system is much smaller than that for a two trophic level system. This makes it less likely that a three trophic level system will persist than a two trophic level system. In short, for finite difference as well as differential equation models there are limitations on the number of trophic levels imposed by population dynamics.

6.5.4 Consequences of dynamical constraints

Is there any evidence for dynamical constraints limiting the number of trophic levels? A potentially testable prediction is that food chains should be shorter in less predictable environments. As discussed in Chapter 3, in stochastic systems there are forces that move densities away from equilibrium (the perturbations) and those that return the densities to equilibrium (which are measured by the return time of the system). Stability involves the balance between these forces favouring the latter. In unpredictable systems, return times must be short and, consequently, short food chains should predominate.

When John Lawton and I first thought of this possibility we suspected that it would remain untested for a long time. Recently, Kitching's (1981) remarkable study of tree-hole communities has come to our attention. This study presents detailed food webs of communities that exist in the water that accumulates in depressions in roots and the branches of trees. The source of energy is provided by leaves that fall into the holes. Kitching studied two communities, one in England and the other in Australia. Both were remarkably similar in surface area, pH, conductivity (a measure of dissolved inorganic salts) and annual litter fall all of which were statistically indistinguishable in the two areas. Only the depth of the holes was different, the Australian holes being, on average, only 65 % as deep as the English ones. Despite these similarities, the Australian community had three trophic levels (detritus, detritivores, predators) and the English community only two (Fig. 6.10). The detritivores in both communities

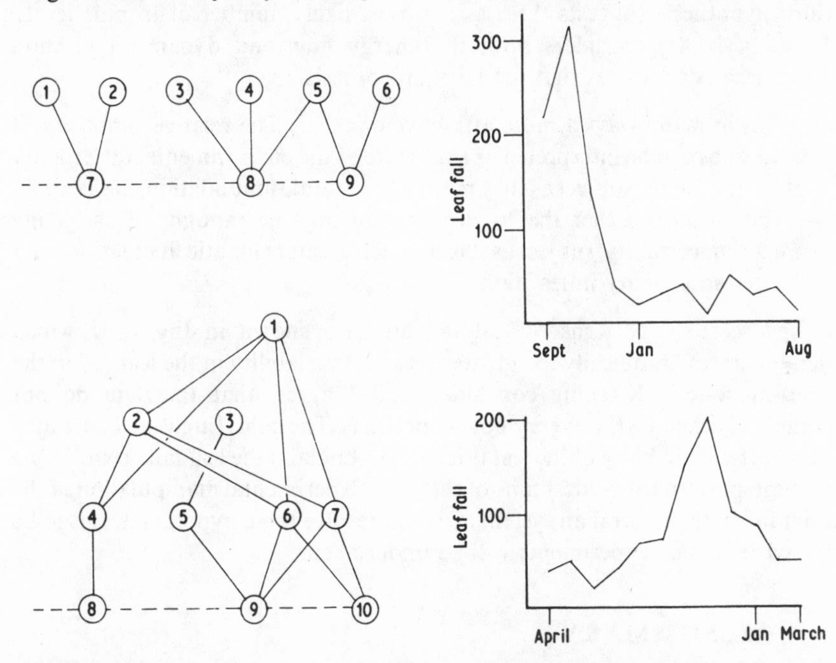

Figure 6.10 Food webs for two tree-hole ecosystems. Bottom, Australia; top England. On the right is the leaf fall per month in grams per m^2. The months are shifted to make the seasons comparable in both figures. Legends: top: (1) *Myiatropa*, Syrphidae, (2) *Prionocyphon*, Helodidae, (3) *Metriocnemus*, Chironomidae, (4) *Dasyhelea* Ceratopogonidae, (5) *Aedes*, and (6) *Anopheles*, both Culicidae, (7), (8) and (9), large and small detritus particles and dissolved organic matter. The broken line indicates that (7), (8), (9) are arbitrary stages along a size-continuum of detritus. Bottom: (1) *Lechriodus*, Leptodactylid frog, (2) *Anatopynia*, Chironomidae, (3) *Cheiroseius*, Ascid mite, (4) *Prionocyphon*, (5) Hyadesiid mite, (6) *Aedes* species 1, and (7) *Aedes* species 2, both Culicidae, (8), (9) and (10), as (7), (8), (9) in top figure. I have omitted two species (*Culicoides angularis*, Ceratopogonidae, and *Arrhenurus*, a prostigmatid mite) that are predatory, but whose trophic relations are uncertain. After Kitching (1981).

are very similar. In the English system, the detritivores are members of the families Chironomidae, Ceratopogonidae, Culicidae, Syrphidae (all Diptera) and Helodidae (Coleoptera). In the Australian system Syrphids are missing, and Chironomids and Ceratopogonids are present but, interestingly, are predators not detritivores. The Australian system also has predatory mites and, at the top of the web, a Leptodactylid frog. The similarity of physical and chemical environments and the taxonomic compositions of the communities suggests that the second and third hypotheses presented above are unlikely explanations for the different numbers of trophic levels. Thus, if design constraints preclude predatory habits for one system they should do so for the other. And it is unclear why food chains should contract to two trophic levels for evolutionary reasons in England, but not in Australia, especially as the third hypothesis suggests three as the most likely number of trophic levels. Kitching (1981) considers both the energy flow and dynamic limitation hypotheses, and writes that the latter hypothesis

> '. . . is in many ways a more attractive one . . . . The relative simplicity of both webs can be interpreted as a measure of the environmental uncertainty of both systems, subject as they may be to periodic innundation, desiccation, or pollution. Further the English system may be thought of as having greater uncertainty, subject as it is to much greater climatic fluctuations and greater seasonality in leaf fall.'

The seasonal patterns of leaf fall are also shown in Fig. 6.10, which demonstrates graphically the greater seasonal variability in the leaf fall in the English system. Kitching concludes, and I agree, that the data do not completely exclude the energy flow hypothesis. The cold, English climate may reduce the availability of the leaf litter and in England the leaf fall occurs in the autumn prior to the coldest time of the year. Experimental manipulation of the litter fall in the Australian system may well resolve these hypotheses. It is to be hoped that such experiments will be undertaken.

6.6 SUMMARY

The number of trophic levels in ecosystems is typically three or four. This chapter seeks explanations for this phenomenon. Four hypotheses are presented and their consequences discussed.

Energy flow

Of the energy consumed by a species, only a fraction goes to make new animals. Most of it is 'lost' as heat and some of it is not assimilated by the animal. The proportion of the energy consumed that is available to the species' predators is called the ecological efficiency. It is small, of the order of 10 %. As energy is transferred along a food chain, that available to any subsequent species becomes drastically reduced. Thus, goes the argument, there comes a point where there is insufficient energy to support another trophic level.

I review the patterns of primary productivities and ecological efficiencies. Primary productivities range over several orders of magnitude and ecological efficiencies vary from less than 3 % in endotherms to nearly 60 % in insects. These observations suggest a number of testable hypotheses.

(i) The greater the primary productivity, the longer could be the food chains. Data from terrestrial and freshwater habitats, however, show that this is not the case.

(ii) There should be some threshold level of primary productivity below which a second and third trophic level cannot be supported. An analysis of the standing crop of the third trophic level against primary productivity shows that at very low levels a third trophic level may be impossible. These levels are at the extreme lower end of primary productivities recorded for natural systems, but the results do confirm that, in the extreme, energy flow imposes limits on the lengths of food chains.

(iii) Because, for a given primary productivity, more energy is available to the consumers of insects than of ectothermic vertebrates than of endothermic vertebrates, food chains should be longer in the first of these cases than the last. They are not.

I conclude that energy cannot be the sole factor in limiting the length of food chains.

Size and other design constraints

A Gyrfalcon may be at the end of a food chain because of the physical impossibility of having a winged predator fast enough to catch the bird and large enough to kill it. Though appealing, such arguments are *ad hoc* and difficult to test. Moreover, the world has been and is full of 'impossible' beasts. Examples include pterosaurs nearly an order of magnitude larger than the largest modern birds and small hawks capable of catching such highly manoeuvrable birds as hummingbirds.

The increased size that accompanies feeding at higher trophic levels means increased energy requirements. These increase approximately as the square of the body length. But increased size also has its benefits. Larger animals feed over larger areas than smaller ones. If energy gain is proportional to an animal's home range then it, too, increases as the square of the animal's length. The cost to benefit ratio does not increase inevitably with trophic position and body size.

Optimal foraging

Because of the patterns of energy transfer, there is potentially more energy available at lower trophic levels than at higher ones. Food chains may be short because of evolutionary trends for animals to feed at the lowest trophic level consistent with dietary requirements. For a particular species, the potentially greater energy available at lower trophic levels may be balanced by the greater number of competitors for that energy as well as by increased susceptibility to predators. There are reasons for feeding low in the food chains, but there are also reasons for feeding high in them. The limitation in the number of trophic levels may be explained by a balance between these opposing processes.

Dynamical reasons

Long food chains are dynamically fragile. Increasing the length of food chains greatly increases their return times. This means that such systems will not likely survive the frequent stochastic perturbations of the real world. The systems that do survive will have shorter return times and, accordingly, shorter food chains. For systems best modelled by finite difference equations, long food chains also imply decreased chances of stability.

I discuss two tree-hole communities in Australia and England. The former has three, the latter, two, trophic levels despite their nearly identical physical conditions. Energy enters the systems as leaf litter and, annually, total leaf fall is similar in both places. But in England leaf fall is less regular than in Australia. These results are compatible with the hypothesis of dynamical limitation of food chain length.

In short, four hypotheses have been presented to explain the short lengths of food chains in the real world. Energy flow restrictions and design constraints seem unlikely causes, though there is no doubt that, in the extreme, energy flow precludes long food chains. Still, it would be premature to select any single hypothesis as the sole explanation for the observed limitation in the lengths of food chains. Clearly, far more data are needed to resolve this problem.

APPENDIX 6A: DRAWING INFERENCES ABOUT FOOD WEB ATTRIBUTES

In this appendix I sketch the methods by which the estimates of the values of P in Table 6.1 were obtained. The methods are general and I use them in several other chapters when I wish to draw inferences about whether a particular food web attribute occurs more or less frequently than chance alone dictates. For most food web attributes there are no known statistical distributions such as the student-t and chi-square distributions that are used so often to test results. The alternative is to generate the statistical distributions on a computer. In performing a statistical test, the probability of an observed result (α) is calculated given that a hypothesis, the null hypothesis, is true. If this probability is sufficiently small (usually < 0.05) then the null hypothesis is rejected. There are practical as well as philosophical reasons why this route is used, rather than the other route of showing that, given the observed result, the other hypothesis (the alternative) is highly probable. First, the null hypothesis is usually a simple one (e.g. the distribution of a food web characteristic is random) and so it is easy to calculate the probability of an observed result given that the null hypothesis is true. Second, data can absolutely reject a hypothesis (e.g. one pine tree rejects the hypothesis that all trees lose all their leaves once a year), but can never certify that a hypothesis is correct (e.g. areas can be found where all the trees encountered are deciduous).

The null hypothesis

There can be many null hypotheses: the more complex the null hypothesis, the more complex the notions rejected when the null hypothesis is shown to be improbable. The null hypotheses I seek to reject are that food webs lack pattern, that is they have the attributes of purely random processes. Without any qualifications this hypothesis is one of 'any species can eat any species'. This leads to food web designs that are so obviously absurd that it would be pointless to spend any effort rejecting such a hypothesis. Even if the statistics failed to reject the null hypothesis, it would still not be *accepted* because of its absurdity. Progress in science, as Lakatos (1978, p. 31) has observed, comes from three-cornered fights between plausible rival hypotheses and the data, not from two-cornered fights between one hypothesis and the data.

To make the null hypothesis plausible, random webs are produced within certain biologically reasonable contraints so that patently absurd webs are eliminated. Then, if the null hypothesis is rejected, the biological conclusion is a more interesting one. In fact, the random webs created under the null hypothesis should resemble real-world webs as closely as possible and differ only in those features about which conclusions are to be drawn. This ensures that variables that have been constrained do not lead to silly results. I shall provide an example of this below. But first, these are the constraints I used: webs had no loops of the kind in which species A eats species A (cannibalism); or species A eats species B eats species A, etc. Such loops result in an indeterminate number of trophic levels as I define them. Examples of such loops in the real world are rare. Further, I restricted the random webs to those that had the following parameters identical to those in the real-world webs:

(i) n, the number of species of prey.
(ii) m, the number of predatory species.
(iii) n_b, the number of species which feed on nothing else within the web. These are usually plants, but they also include species which feed on detritus and other heterotrophs whose prey are not specified. I call these basal species.
(iv) n_p, the number of species on which nothing else in the web feeds (top-predators). Note that $n + n_p = m + n_b =$ the total number of species.
(v) n_{int}, the number of predator–prey interactions in the web.

Restricting the webs in the ways described minimizes the number of trophic levels found in the random webs, thus making the null hypothesis less likely to be rejected. As an example, suppose that only $n + n_p$ (the total number of species in the system) was fixed. Up to $n + n_p$ trophic levels would be possible.

Of course, such a null hypothesis might be more easily rejected, but its rejection would be less interesting: the null hypothesis would produce absurd food webs with herbivorous plants feeding on other herbivorous plants and so on.

The data

To test the hypothesis that food chain lengths are limited, a large set of systems is chosen from as wide a range of habitats as possible. For each system information must be available on the choices of prey for each species within the system. Given this information, it is a simple matter theoretically to calculate the numbers of trophic levels and to compare them to the model systems based upon them. Unfortunately, the assessments of which species feed upon which other species are usually subjective and the documentation of the webs varies considerably. This subjectivity is not necessarily a problem since the data are to be treated statistically. I shall be concerned with statistics averaged over all the data and much of the subjectivity in producing webs will be reflected only as experimental error. Other subjective assessments may be reflected as bias, so webs must be selected that will minimize bias.

Quite independently (Pimm, 1980b), I excluded many studies for the same reasons as Cohen (1978), but there are some studies I used that Cohen omitted and vice versa. I excluded some webs used by Cohen on three grounds:

(i) Some webs had all the species in one of two trophic levels.

(ii) In some studies the entries to the webs were potentially heterogeneous groupings of species (e.g. 'caterpillars' or 'large ground animals').

(iii) Some webs were too large to be analysed with existing computer facilities. These webs do not contain high numbers of trophic levels. Yet the large random webs are capable of having a large number of trophic levels. Indeed, there is a negative correlation between the size of the web (as measured by the number of predators it contains) and the value of P in Table 6.1. This suggests that were these large webs to be analysed they would lead to an even more convincing rejection of the null hypothesis.

Unlike Cohen, I included some studies on man-modified systems. In the next chapter I shall compare food webs dominated by insects with those dominated by vertebrates. The former are from largely agricultural systems, so it seems most consistent to include all data from man-modified systems.

Finally, there are studies that both Cohen and I overlooked at the time our work was done. Some of these appear elsewhere in this book, but as none bristle with long food chains their omission is not likely to have affected the results.

A number of the studies included quantitative information on the diet of the species. In all of these, considerable complexity arose when an interaction was indicated between two species yet the prey accounted for only a trivial proportion of the predator's diet. Some webs (indicated by S in the table) have been simplified by excluding interactions between predator and prey up to the point where 5% of the predator's total food intake has been accounted for. This probably excludes interactions whose strength is so small as not to affect seriously the stability of the system. Of course it is quite possible that an item occurring only rarely in the diet might be the single most important factor limiting population size. The long list of trace elements required for proper human nutrition comes immediately to mind, but all such elements might not occur in the commonest prey. Only three webs have been simplified, however, and their exclusion does not alter the results. None the less, the problem of what constitutes the controlling interactions within a web remains and will do so until further field work is undertaken.

Remaining biases

Despite these attempts to minimize the biases in the data, some will remain. At least two processes may be involved.

(i) Some webs may be so complex that this precludes drawing simple diagrams of them. Even if such diagrams could be produced, the system's complexity may inhibit the ecologist from studying the system in the first place. After all, scientists usually seek simple patterns and the hopefully simple processes that underlie them. When, in subsequent chapters, I conclude that webs are relatively simple, there is no way I can exclude for certain the hypothesis that only simple food webs are published. The nearest I can come to defending the data is to anticipate results of later chapters that show that some long-term and highly detailed studies of food webs yield qualitatively the same patterns as more extensive and inevitably less carefully conducted studies that constitute the bulk of the data. Of course, such carefully conducted studies cover only a limited number of ecosystems, and this raises the second possible source of bias.

(ii) All kinds of food webs are not equally represented in the data: freshwater and intertidal systems are common; soil ecosystems (for example) are absent. However, the available data are usually homogeneous across a wide range of different ecosystems. Food chains, for example, are not noticeably of different lengths in terrestrial, freshwater and marine systems (Table 6.1). None the less, some studies of systems not included here may uncover webs with unusual properties. Such results will demand the asking of the question: why do systems differ? For the present it is sufficient to ask why many systems, at least at a gross level, share so many properties.

The expected distributions of food web characteristics

(a) Synthesis of food webs

Consider a rectangular matrix, Q, whose typical entry q_{ij} is 1 if j feeds on i, 0 otherwise. Prey species (i) are numbered from $n_p + 1$ to $n + n_p$ and predator species (j) are numbered from 1 through m. Species numbered 1 through n_p are top-predators and species numbered $n+1$ to n_p+n are basal species. Species that are neither basal nor top-predators obviously appear as both prey and predators.

I created model Q matrices by placing n_{int} interactions into a matrix numbered as described. The constraint of no loops described above requires that

$$q_{ij} = 0, \quad \text{for all } j \geqslant i.$$

Next, the matrices were checked to see that each prey had a predator and each predator a prey. That is,

$$\sum_{j=1}^{m} q_{ij} \geqslant 1 \quad \text{for all } j$$

and

$$\sum_{i=n_p+1}^{n+n_p} q_{ij} \geqslant 1 \quad \text{for all } i.$$

If Q failed these tests, I chose another randomization.

(b) Analysis of model food webs

The computer program counted each pathway from each top-predator to all basal species from which the top-predator received some energy. The modal path length plus one was returned as the modal number of trophic levels. When there was more than one top-predator in each web, a vector of modal trophic levels was returned. The program compared the vector of the modal trophic levels to the vector of modal trophic levels in the real web. When the vectors were different, the program had to decide whether the model web had more or fewer trophic levels than the real web. No problem was encountered if there was a single top-predator, nor was it encountered in most other circumstances when there were multiple top-predators. However, when the real web had, say, two top-predators associated with three trophic levels each, and the model web had two and four, there was no clear answer. I wrote the program to choose the model web as having fewer trophic levels in such circumstances. This decision maximizes the proportion of model webs that are simpler than the real web (P) and is, therefore, conservative.

To illustrate the process, suppose that the web shown in Fig. 6.11 was the real web. It would have parameters $n = 5$, $m = 5$, $n_p = 2$, $n_b = 2$ and $n_{\text{int}} = 7$. The modal number of trophic levels would be three (for X_1) and four (for X_2). The corresponding matrix, Q, is shown in Table 6.3. To examine how improbable is such a structure, a large number of matrices must be created with these parameters. For explanation, three examples (Table 6.3, b–d) will suffice.

The first example (b) contains a loop (X_4 feeds on X_5 which feeds on X_4) and is excluded.

The second example (c) does not have a loop. The first top-predator to be

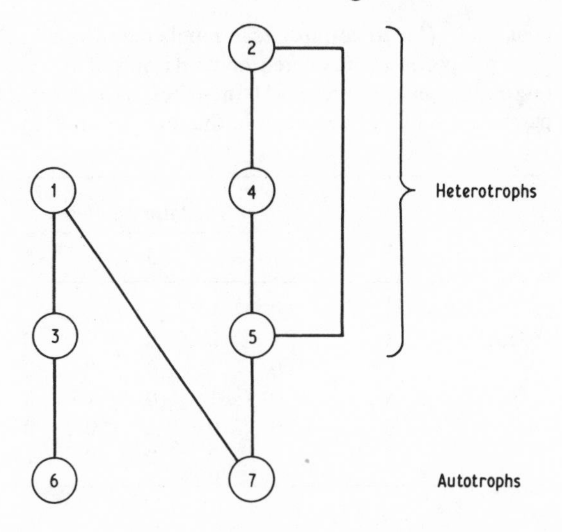

Figure 6.11 A food web used to illustrate the methods used to generate expected food web characteristics (see text). (From Pimm, 1980b.)

analysed is X_1. There are two pathways from it to the basal species; one is to X_6 via X_3, giving a value of three for the number of trophic levels. The second pathway is to X_7 via X_3, X_4 and X_5, giving five as the value for the number of trophic levels. Since the values of three and five occur equally frequently (once) the program would choose five as the modal number of trophic levels. This is the same convention used to calculate the modal number of trophic levels of real webs. Although arbitrary, the decisions are consistent. The second predator to be analysed is X_2. Connections from it to the basal species in three ways: one to X_6, and the other two to X_7. The three values for the number of trophic levels returned would be three (to X_6, via X_3), two (to X_7 directly) and the five (to X_7, via X_3, X_4, X_5). The analysis would choose five as the modal number of trophic levels for both top-predators. This web would be considered among the set that had more trophic levels than the real web.

The third example (d) is similar to that of (a) (and Fig. 6.11) except that X_1 feeds on X_6, not X_7. However, the number of modal trophic levels is unaltered and this web would be placed in the set of model webs that have as many, or fewer, trophic levels as the real web.

The final stage is simple. The proportion P is calculated as the number of model webs that have trophic levels less than or equal to the real web, divided by the number of webs generated by the analysis. Inferences should be drawn about a set of values of P because the null hypothesis relates to food webs in general, rather than a specific web. This can be done using the chi-square test on proportions described by Fisher (1950). It involves taking twice the natural

Table 6.3 Q matrices: species in numbered rows are the prey to species in numbered columns only if the corresponding entries are 1. All matrices have identical parameter values as described in the text. (From Pimm, 1980b.)

		Predator species				
		1	2	3	4	5
a						
Prey species	3	1	0	0	0	0
	4	0	1	0	0	0
	5	0	1	0	1	0
	6	0	0	1	0	0
	7	1	0	0	0	1
b						
	3	1	0	0	0	0
	4	0	1	0	0	1
	5	0	0	0	1	0
	6	1	0	1	0	0
	7	0	0	0	0	1
c						
	3	1	1	0	0	0
	4	0	0	1	0	0
	5	0	0	0	1	0
	6	0	0	1	0	0
	7	0	1	0	0	1
d						
	3	1	0	0	0	0
	4	0	1	0	0	0
	5	0	1	0	1	0
	6	1	0	1	0	0
	7	0	0	0	0	1

logarithm of the proportion, which gives a chi-square value with two degrees of freedom. Chi-square values are additive, as are their degrees of freedom, so a probability of the null hypothesis can be obtained for all the webs analysed. It is this value that is used to decide whether the webs are different from those produced under the 'random but constrained' hypothesis.

7 The patterns of omnivory

When discussing food webs it is convenient to define an omnivore as a species that feeds on more than one trophic level (Pimm & Lawton, 1978). In some food webs, there are few omnivores (Fig. 7.1 (a)), while in others omnivores are common (Fig. 7.1(b)). This chapter examines patterns of omnivory using the method outlined earlier, that is, real webs are predicted to have those structures that enhance stability in models. This method gives several predictions that can be tested against real food web structures. These predictions are supported by the data, and this constitutes the most certain evidence that dynamics impose constraints on food web structures. Moreover, the success in predicting specific patterns of omnivory provides the least equivocal evidence that increasing complexity decreases the likelihood of stability in the real world. In Chapter 5, I argued that the relationship of

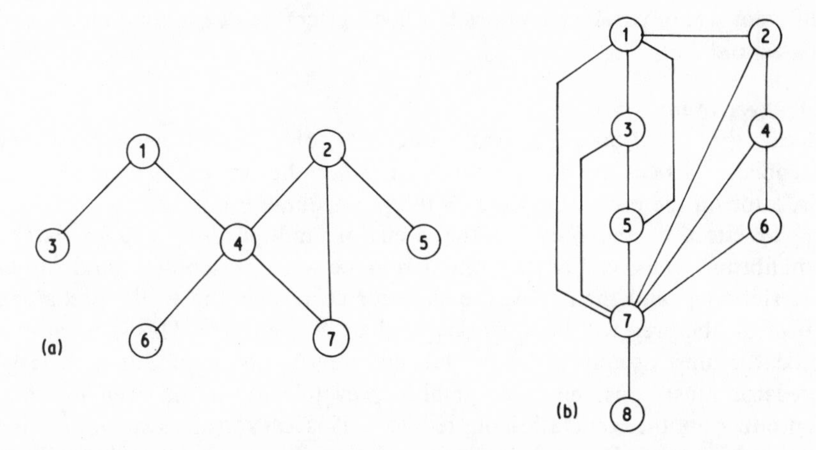

Figure 7.1 Two food webs. In (a) there is only one omnivore, in (b) omnivory is common. (a) After Paine (1963), represents the interactions between intertidal gastropods, pelecypods, and their prey; (b) After Force (1974) represents the interactions between *Bacharis*, its herbivorous insects, and their parasitoids. This system has a loop: species 1 eats species 2, and vice versa. Legend: (a) (1) *Pleuroploca gigantea* (2) *Fasciolaria tulipa* (3) *Atrina rigida* (4) *Fasciolaria hunteri, Busycon contrarium, Murex florifer* (5) *Conus floridana, Turbo casteneus* (6) *Cardita floridana* (7) *Chione cancellata*, various pelecypods and polychaetes (b) (1) *Amblymerus* (2) *Zatropis* (3) *Torymus koebelei* (4) *Torymus bacchacicidie* (5) *Platygaster* (6) *Tetrastichus* (7) *Rhapalomyia* (8) *Baccharis* (From Pimm & Lawton, 1978.)

complexity to species number, which is suggested by dynamical considerations, could as easily be explained by simple biological assumptions. Now, other things being equal, increased omnivory means increased complexity. If omnivory is constrained by dynamics, then so is complexity. Unlike the patterns of complexity, the more specific patterns of omnivory cannot be so easily related to simple biological assumptions. Only dynamical constraints predict the detailed patterns of omnivory observed in nature.

7.1 MODELS OF OMNIVORY

7.1.1 Lotka–Volterra models

For models with four species and four trophic levels there are eight possible model configurations (Fig. 7.2). The eight models are distinguished by (i) the *rank of omnivory*, which is the number of predator–prey interactions in excess of those in the model with no omnivory, and (ii) the position of those interactions. Lawton and I used the procedure of establishing matrices with sign configurations corresponding to these models. We then analysed the matrices to determine their maximum eigenvalues and thus their stability and return times (Pimm & Lawton, 1978). The elements of the matrices were chosen uniformly over specified intervals with 1000 replications. We investigated three different sets of parameter values labelled 'vertebrate', 'vertebrate/insect' and 'parasitoid'.

(i) Vertebrate

Recall that for Lotka–Volterra models the off-diagonal elements of the Jacobian matrices are the products of *either* the *per capita* effect of the predator on the prey multiplied by the prey's equilibrium density *or* the *per capita* effect of the prey on the predator multiplied by the predator's equilibrium density. For the interaction between vertebrates (and larger invertebrates) and their prey, the elements corresponding to the predator's effect on the prey will be much larger than the converse. This is because a predator must usually be larger than its prey if it is to subdue it. A large predator must consume many smaller prey to survive and even more to reproduce another generation of predators. This leads to an asymmetry in the *per capita* effects. Moreover, the predator's equilibrium density will usually be smaller than that of its prey. For the predator's effects, we chose parameters randomly over the interval $(-10, 0)$ and for the prey effects, parameters similarly over the interval $(0, +0.1)$. These were the parameter limits I used to investigate the limitations on complexity and food chain length discussed in previous chapters.

(ii) Vertebrate/insect

The *per capita* effect of an insect feeding on a tree is very small, while the converse is very large. Yet, while the *per capita* effect of a bird feeding on the

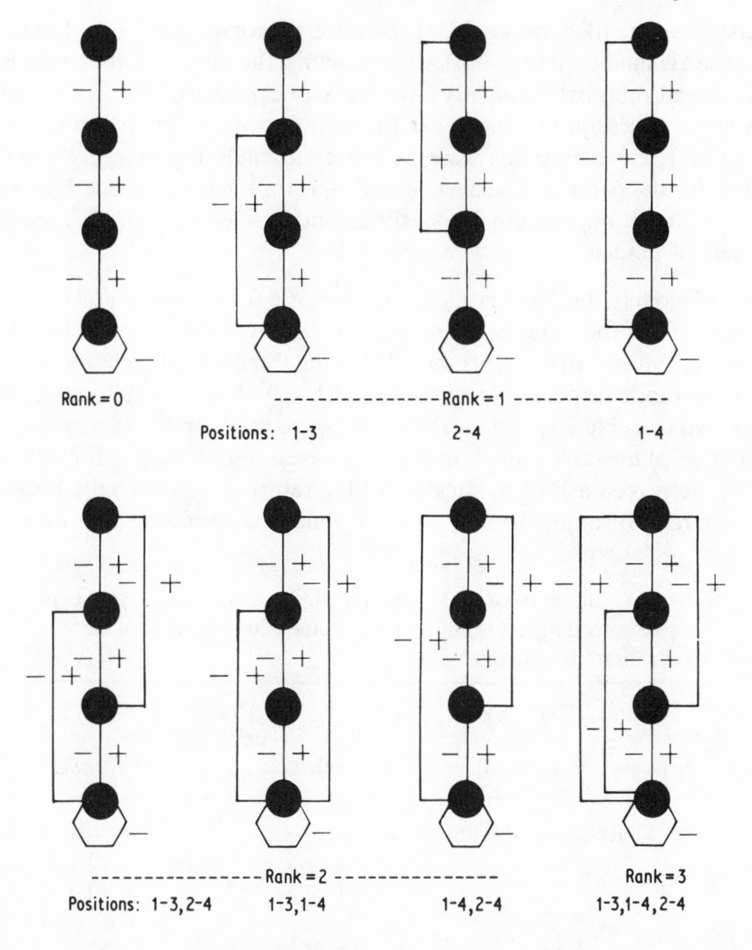

Figure 7.2 The eight possible four-species, four trophic level models to indicate extent (0, 1, 2, 3) and position (1–3, 1–4, 2–4) of omnivory. For sign conventions see Fig. 6.6 (From Pimm & Lawton, 1978.)

insect is large, the reverse is small. The insect is 'sandwiched' between its prey and predator populations, both of which exert a considerable *per capita* effect upon it. For this set of simulations, we chose the effect of the lowest trophic level on the second trophic level to be over the interval $(0, +10)$, the reverse effect over the interval $(-0.1, 0)$, and the other interactions as in (i).

(*iii*) *Parasitoid*

Here, interactions between trophic levels one and two were as in (ii). The remainder were chosen to mimic insect host–parasitoid interactions. In these, the equilibrium population sizes of host and parasitoid may be far more similar than in (i) because the species are of comparable size. Often the parasitoid is smaller than its host. Consequently, the *per capita* effects are more symmetrical: one host frequently gives rise to one parasitoid and, following the heavy mortalities due to the weather, the average parasitoid may capture only a few hosts. These interactions were therefore chosen over symmetrical intervals $(-1, 0)$ for the parasitoid's effect, and $(0, +1)$ for the host's effect. Hence, the average interaction strength of parasitoid models is less than that for the other two sets of models.

For all models, the diagonal elements were zero for all species except those at the base of the food chain. These were chosen over the interval $(-1, 0)$.

The percentages of unstable models for the three sets of eight models and, for stable models, the percentages of models with long (>100) return times are shown in Table 7.1. Note first that, as the rank of omnivory increases, the percentage of unstable models increases. Interestingly, for the subsets that are stable, the percentage of models with long return times generally decreases with the rank of omnivory – (a topic to which I shall return in Chapter 10).

Table 7.1 Stability of 1000 random food webs of each type, constructed according to the range of parameter values specified in the text. (From Pimm & Lawton, 1978.)

Rank of omnivory	*Position (species linked)*	*Models*		
		Vertebrate	*Vertebrate/ insect*	*Parasitoid*
Percentages of unstable models				
0		0	0	0
1	1–3	81	78	40
	2–4	91	91	68
	1–4	96	96	51
2	1–4 & 2–4	98	98	59
	1–3 & 2–4	97	97	74
	1–4 & 1–3	99	99	50
3	1–3, 1–4 & 2–4	100	100	72
Percentages of stable models with return times >100				
0	–	44	41	42
1	1–3	21	22	19
	2–4	11	13	19
	1–4	47	46	38
2	1–4 &2–4	–	–	11
	1–3 & 2–4	–	–	13
	1–4 & 1–3	–	–	13
3	1–3, 2–4, 1–4	–	–	13

Percentages are given to the nearest integer.

Thus, the effects of stability and short return times work in opposition to one another. In the real world, the likelihood of finding a web might increase or decrease with the extent of omnivory. It might decrease because the fraction of stable models decreases with increasing omnivory, or increase because only models with short return times can persist (return times decrease with increasing omnivory). Whereas return times shorten only slightly with increasing omnivory, for vertebrate and vertebrate/insect models, such a small fraction of models were stable for omnivory of rank greater than one that such structures are unlikely to be found. The likelihood of stability must surely decrease as the extent of omnivory increases. Increasing omnivory also increases complexity and, as discussed earlier, this rapidly diminishes the chances of a model being stable. This result is quite general. For difference equation models of insect host–parasitoid systems, omnivory also places additional restrictions on stability (May & Hassell, 1981).

Next, models with only one omnivore differ in their stability depending on how many trophic levels are spanned by the omnivore. For vertebrate models, 19 % are stable when the carnivore feeds on both the herbivore and the plant, 9 % are stable when the top-carnivore feeds on the herbivore and the carnivore, but only 4 % are stable when the top-carnivore feeds on the carnivore and the plant. Moreover, for this third group, the return times are generally much longer and reverse the trend of shorter return times with smaller percentages of stable models.

Finally, although vertebrate and vertebrate/insect models show almost identical patterns, the omnivorous parasitoid models were much more likely to be stable. The return times for the three sets of models were generally similar.

This comparison of parasitoid models with the other two systems is based on continuous differential equations, rather than on discrete difference equations which are probably better describers of most insect populations. Difference equations place more restrictive conditions on stability than differential equations. (Recall that the former require only the real part of the eigenvalues to be strictly negative; the latter require all the eigenvalues to lie in a unit circle, centered on the origin, in the complex plane.) The greater potential complexity of insect systems should be viewed as only an upper bound. Extensive omnivory may be possible for some insect systems, but will obviously be reduced by the additional stability restrictions imposed by discrete life histories. May & Hassell (1981), however, have also pointed to the process of *pseudo-interference* (first discussed by Beddington, Free & Lawton, 1978) as another factor adding to the potential robustness of complex insect systems. In pseudo-interference, insect predators/parasitoids exploit highly clumped prey in a density-dependent fashion that adds greatly to the stability of the predator–prey interaction. If insect-dominated webs are shown to reach complexities greater than 'vertebrate' webs, then pseudo-interference and/or the lower interaction strengths between hosts and their parasitoids may be the cause.

These results are based on simple models with no more than four species. I have investigated more complex models with six and eight species organized into two chains (see Fig. 4.6) and obtained the same results (Pimm, 1979(b)). In the six- and eight-species models, extensive omnivory destabilized webs. Omnivory involving prey species not at adjacent trophic levels, destabilized models more so than when omnivores fed on prey in adjacent trophic levels. The latter group have long return times and are exceptions to the trend for return times to decrease as the percentage of stable models decreases. Some of these data (on the percentage of stable models for eight-species systems) are presented in Table 8.1 where they are discussed in a different context.

In short, three predictions emerge from these results:

(i) omnivory should be relatively rare;

(ii) omnivores should only rarely feed on species in trophic levels that are not adjacent; and

(iii) food webs composed of insects and their parasitoids may have much more complex patterns of omnivory than the webs dominated by vertebrates and their prey.

7.1.2 Donor-controlled models

I have argued that donor-controlled dynamics do not describe the majority of interactions between predators and their living prey, but that they are likely to be the better description of the interactions between detritivores and their dead food. In not exerting control over the growth rates of their food, such species do not destabilize systems with extensive omnivory.

Consider the food web illustrated in Fig. 7.3. Species X_5 feeds on all four species (and trophic levels) but has no effect on them; the species X_1 through X_4 are arranged in a simple food chain (say, plant, herbivore, carnivore, top-carnivore). X_5 is a detritivore that feeds on dead animals and plants, and it is highly omnivorous. To check the stability of this model use the criteria required for qualitative stability on page 30: (a) Condition (i) is satisfied because X_1 is self-limited. (b) The system fails the colour test for the same reason as did the web in Fig. 2.3(c): the links to species X_5 do not bring it within the predation community of the other four species because this species does not have a negative effect on the growth rates of those species. (c) There are no mutualistic or competitive interactions. (d) There are no *p* cycles longer than two. (e) The determinant is not zero. Despite the complex patterns of omnivory, the system is qualitatively stable. Of course, if X_5 affects the density of any of its prey then *p* cycles of three or longer are formed and the system is no longer qualitatively stable. Compare this system, with no unstable models, with the four-species Lotka–Volterra model with three omnivore links of Table 7.1 and Fig. 7.2, which has fewer omnivores but where fewer than 5 models out of 1000 are stable. The assumption of whether a predator affects

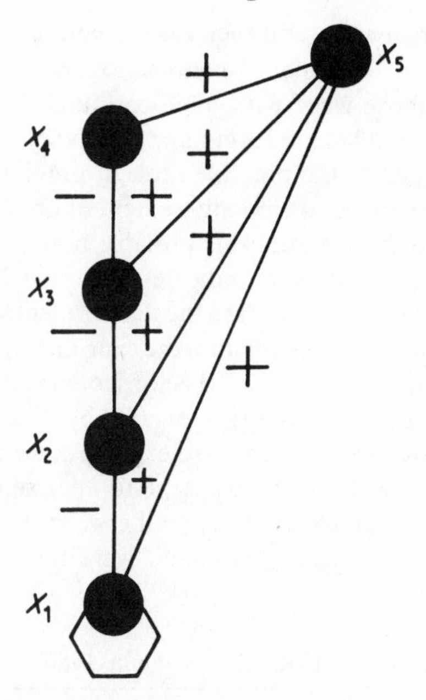

Figure 7.3 A digraph of a qualitatively stable model with extensive omnivory. Note that species X_5 has no effect on the growth rates of its 'prey'. For other sign conventions see Fig. 6.6.

the density of its prey is clearly crucial in determining which patterns of omnivory are likely.

In short, there is a fourth prediction:

(iv) omnivory can be extensive if it involves species that do not affect the population growth rates of their prey.

7.2 TESTING THE HYPOTHESES

For hypothesis 1 to be correct, real food webs must have fewer omnivores than webs generated at random. Again, it is not known in advance how the extent of omnivory should be distributed, for there are no obvious statistical distributions to use. The distribution of omnivory under the null hypothesis (that omnivory is random) must be obtained by generating random webs on a computer. The problem is directly comparable to that of deciding whether food chains are shorter than expected.

The methods used were the same: the computer program created a sample of random webs with the same number of species of prey, predator, basal species and top-predator and the same predator–prey interactions as the real web. From these it selected only those webs that had precisely the same number of trophic levels as the real webs, and then calculated the proportion P that had omnivory to an extent less than, or equal to, that of the real web. The program used only those webs with identical numbers of trophic levels to separate the effects of food chain length and omnivory. In general, the longer the food chain, the greater the possible rank of omnivory. Food chains are generally short and so restrict the possible extent of omnivory. The restriction of identical chain lengths excludes this possible bias.

A sufficiently small value of P indicates that real food webs have less omnivory than expected by chance (see the data presented in Table 7.2). Were real food webs to have random omnivory, the expected value of P would be 0.5; the observed mean value of P is 0.253. The observed value is significantly less than the expected value ($\alpha < 0.01$) (Pimm, 1980b).

In Appendix 6A I discussed the need to exclude some webs where the large number of species made the webs too large to analyse on the computer. In Table 7.2 there is a negative correlation between P and the number of predators in the web, which suggests that, were these large systems to be

Table 7.2 Omnivory statistics. (From Pimm, 1980b.)

P*	$\sum R/n_p$[†]	*Type of system*‡	*Source*
(a)	1.0	prairie	Bird (1930)
0.20	0.25	willow forest	Bird (1930)
0.57	3.00	starfish	Paine (1966)
0.12	2.60	freshwater stream	Minshall (1967)
0.66	0.50	dung	Valiela (1969)
0.61	2.00	marine fish	Parsons & Lebrasseur (1970)
0.03	2.00	freshwater pond	Hurlbert *et al.* (1972)
0.00	0.00	mudflat	Milne & Dunnett (1972)
0.07	0.00	mussel bed	Milne & Dunnett (1972)
0.02	0.50	freshwater stream	Jones (1959)
(a)	0.25	freshwater spring	Tilly (1968)
0.25	0.33	lake fish	Zaret & Paine (1973)
(b)	0.00	lake fish	Zaret & Paine (1973)

* The proportion of random webs that have less omnivory than the real web on which they are based. The sample of random webs has the same number of predators, prey, top-predators, basal species, interactions, and modal trophic levels as the real web.

† The number of omnivore interactions which if removed would result in no omnivory, divided by the number of top predators (n_p) in the web.

‡ Additional parameters for these webs are given in Table 6.1.

(a) Insufficient samples were available from the computer analysis to calculate a value.

(b) There is only one way to shape this web when the parameters listed under * are fixed.

included, the mean P would be even smaller and the null hypothesis would have been even more soundly rejected.

In short, the first prediction is supported. Food webs have omnivores – an average of about one per top-predator – but there are fewer of the patterns of extensive omnivory than would be expected by chance alone. Only rarely are two or more omnivores per top-predator observed.

Testing the second prediction is easier. Lawton and I used data from Table 7.2 and from additional webs from Cohen (1978) that were too large for the computer to use to test hypothesis 1 (Pimm & Lawton, 1978). A total of 23 food chains, with one omnivore per chain, were obtained from these webs. Of these 23 food chains, only one had an omnivore feeding on species that were not at adjacent trophic levels. With only one omnivore per chain and with four trophic levels, there should be twice as many species feeding on adjacent trophic levels as feeding on non-adjacent levels (Fig. 7.2: rank1; positions 1–3 and 2–4 versus 1–4). The expected numbers would be: feeding on adjacent levels $(2/3 \times 23) = 15.33$; not feeding on adjacent levels $(1/3 \times 23) = 7.67$. The observed values, 22 and 1, respectively, lead to a chi-square value of 9.63. The null hypothesis is rejected ($\alpha < 0.01$). Omnivores feeding on species not at adjacent trophic levels are rarer than would be expected.

These data also provide a test of an alternative hypothesis for the rarity of omnivory. Omnivores may be few because of the dissimilarity of their prey; it may be difficult to exploit prey species at two different trophic levels effectively. The bill of an insect-eating bird is very different from that of a seed-eating species, so neither species is likely to be effective at feeding on the other's prey. Such features must reduce omnivory. But if this alone were responsible for the patterns of omnivory, then it should reduce the occurrence of omnivores that feed on animals and plants at the expense of those that feed on animals at different trophic levels, since the dissimilarity in diet is greater for the former than the latter. This prediction is not supported. Of the 22 omnivores feeding at adjacent levels, those feeding on animals and plants were as common as those feeding on animals at different levels (11 each). Thus, dissimilarity in diet probably contributes to the overall scarcity of omnivory but does not seem to account for the detailed occurrence of the various patterns of omnivory. In short, the data reject the alternative hypothesis that patterns of omnivory are caused by design constraints, but are consistent with the hypothesis that dynamics restrict the patterns of omnivory.

For the third prediction, namely that insect-dominated webs are more complex than others, I calculated the mean number of omnivores per top-predator for each food web in Table 7.2 and for a set of insect-dominated webs in Table 7.3 (Pimm, 1980b). For insect-dominated webs, there is an average of 2.5 omnivores per top-predator; for the others the average is 0.96. The difference is statistically significant ($\alpha < 0.01$) and the third prediction is supported.

I am indebted to Richard Haedrich (Haedrich & Rowe 1977, pers. comm.) for

Table 7.3 Characteristics of omnivory in systems dominated by insects and their parasitoids and predators. (From Pimm, 1980b.)

n*	m	n_p	n_b	n_{int}	Tl_{mod}	P	$\Sigma R/n_p$	*Type of system*	*Source*
9	10	2	1	15	3	0.133	1.00	collards	Root (1973)
4	4	1	1	6	5	(a)	2.00	winter moth	Kowlaski (1977)
7	7	2	2	15	4	0.69	4.00	oak galls	Askew (1971)
8	8	(b)	1	17	–	–	3.00	coyote bush	Force (1974) Fig. 7.1(b)

* n is the number of prey, m the number of predators, n_p the number of top-predators, n_b the number of basal species, n_{int} the number of interations, Tl_{mod} the modal number of trophic levels, P the proportion of random models that have less omnivory than the real web on which they are based, and $\Sigma R/n_p$ the number of omnivore interactions that would need to be removed to eliminate all omnivory, divided by the number of top-predators.
(a) There is only one way to arrange this web when all the parameters (see Appendix 6A and text) are fixed.
(b) This web contains a loop of the kind A eats B eats A. This loop was omitted in calculating the extent of omnivory.

data relating to the fourth prediction: that omnivory should be common among detritivores. He describes a deep-water fish (*Coryphaenoides armatus*) which attains relatively high densities by feeding on 'fast-falling . . . bodies of fish, whales, squids and decapods'. The fish feeds on a wide range of prey from several trophic levels. There seems little doubt that this system is donor-controlled because the fish feeds on the carcasses after they have sunk to the bottom of the ocean. I suspect that many terrestrial detritivores also feed on a variety of trophic levels. I have seen Turkey Vultures (*Cathartes aura*) take a wide variety of dead herbivores (from as large as donkeys to as small as doves and rodents) as well as carnivores from a variety of trophic levels (snakes, coyotes, hawks and insectivorous mammals). These observations are consistent with the fourth prediction but there are certainly other, simpler explanations. While it requires radically different adaptations to take a bite from a living small rodent instead of its living, avian predator, their carcasses must offer little difference. The dermestid beetles that attack dead organisms are probably as efficient at disposing of a small herbivorous mammal as that mammal's avian predator. The extensive omnivory among detritivores may as well be due to the easing of design restrictions involved in feeding on animals at different trophic levels as to the easing of the dynamical constraints which apply when donor-controlled dynamics obtain.

Finally, I must ask whether these results might be simply reflections on the biases of the authors who collected the data. One procedure to evaluate this possibility was discussed briefly in Appendix 6A ('Remaining biases'). It is to examine those studies that are clearly the most detailed and carefully presented and to ask: do these differ from the remainder of the data? Thus, we should ask: do the scarcity of omnivores reflect the authors' reluctance to publish messy-looking food webs? While this will always remain a possibility until the set of detailed food web studies is considerably expanded, there is no evidence of its validity at present. Paine's detailed studies of intertidal systems (Paine, 1963, 1966, 1971, 1979, 1980) and Jones' quantitative studies of freshwater stream (Jones, 1959) are not unusually laden with omnivores. In contrast, those long term studies of insect systems (Askew, 1971, Force, 1974) are clearly of much greater complexity.

In short, while biases towards simple food webs will remain until further studies are conducted to specifically test the ideas in this (and other) chapters, there is no evidence at present that detailed studies show patterns different from more superficial studies.

7.3 SUMMARY

When model food webs are required to be dynamically stable, four predictions about omnivory can be made: (i) omnivory should not be extensive; (ii) omnivores should not feed on species that are not at adjacent trophic levels; (iii) insect-dominated systems may achieve much greater complexity in their patterns of omnivory; and (iv) systems best modelled by donor-controlled dynamics may also achieve greater complexity. All four hypotheses are supported by data, though the last is perhaps better explained by the idea that plants and animals from any trophic level offer much the same challenges to the consumer when dead. At least one alternative explanation for the patterns of omnivory, that dissimilarities in diet preclude feeding on more than one trophic level, fails to explain all of the patterns observed.

7.3.1 Complexity and stability revisited

These results are closely related to those on the complexity of food webs. Other things being equal, increased omnivory means increased complexity, and both lead to a decreased chance of a model being stable. Insect-dominated webs were modelled with lower interaction strength and, as predicted by Equations (4.2a, b), were more likely to be stable than the other webs for a given number of species and connectance. In contrast, webs where interactions are donor-controlled are usually qualitatively stable and can attain considerable complexity. In short, I conclude that, except for donor-controlled systems, stability restricts omnivory and, as a consequence, complexity. Yet, at the end of Chapter 5, I concluded that the data presented there did *not* provide unequivocal support for the idea that stability restricts complexity. The argument was made because the observed hyperbolic relationship between connectance and the number of species in the system had another and simpler explanation which I could not exclude. The different conclusions, here and in Chapter 5, may seem incongruous in view of the similarity of the data from which they were drawn and the relationships of the underlying theory. Of the predictions about omnivory, only one does not follow directly from Equations (4.2a, b) and the results on donor control. This exception is the rarity of omnivores which feed on non-adjacent trophic levels.

The result that stability restricts the design of food webs (of which complexity is one attribute) is an important one. Moreover, it is not a result that seems inevitably true. Thus, I am prejudiced against the stability–complexity hypothesis when alternatives are available. Others disagree (e.g. Rejmanek & Stary, 1979) and it is a matter of taste. The results presented in this chapter, however, have not been explained by factors other than dynamics, and are therefore more certain evidence that stability places restrictions on the possible designs of food webs and, by extension, complexity.

8 Compartments

In previous chapters I made several predictions about simple food webs that involved small numbers of interacting species. This chapter considers the problem of ecosystem design on a larger scale. How are food webs involving a large number of species organized?

John Lawton and I examined two extreme hypotheses (Pimm & Lawton, 1980). Given the constraints on food chain length and the extent and position of omnivory already discussed:

(i) Under the 'reticulate hypothesis', species interactions will be homogeneous throughout the web.

(ii) Under the 'compartmented hypothesis', only species within a particular compartment will interact. Between compartments (called 'blocks' by May, 1972) there will be little or no interaction.

The two hypotheses are illustrated in Fig. 8.1. Actual food webs are unlikely to be as distinct as the two hypothetical extremes shown, but this should not preclude asking which hypothesis provides the better description of the real

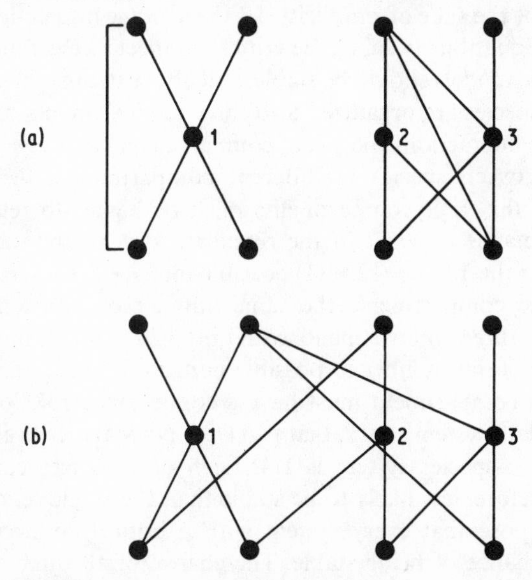

Figure 8.1 Two webs with identical numbers of species, connectance, top-predators, basal species and omnivores. (a) is compartmented; (b) is not. (From Pimm & Lawton 1980.)

world. The alternative, intermediate to these hypotheses, is that interactions are randomly distributed throughout the web. These hypotheses are analogous to the spatial distribution of objects which may be clumped, random or over-dispersed. Furthermore, features of spatial distributions change with the scale over which the objects are observed, for example, deciduous trees may be random or over-dispersed within a forest but are clumped into a few areas in the world. Patterns of compartments within ecosystems may also be a function of scale.

8.1 REASONS FOR A COMPARTMENTED DESIGN

A compartmented design might prevail for at least two reasons. The first explanation comes from dynamical arguments similar to those discussed in previous chapters. I shall present these arguments first. The second explanation involves some simple biological constraints and is discussed in the next section.

8.1.1 Dynamical reasons

May (1972, 1973b) has argued, from considerations of connectance and the number of species in a system, that, for the same average interaction strength, models will be more likely to be stable if their interactions are arranged into compartments. This argument appears to be wrong. Consider a twelve-species reticulate system structured to meet the assumptions leading to Equations (4.2a, b) and, for the sake of simplicity, let the average interaction strength be 1.0. Applying Equations (4.2a, b), the critical connectance is found to be 1/12; below this value models should be stable, and above it, unstable. On the other hand, what if species are organized into three compartments of four species, each with no interactions between compartments? Because of the zero interactions between species in different compartments, the connectance within each of the three compartments must be higher to retain the web's average connectance of 1/12. In the reticulate system, the interactions are distributed over the 132 ($= 12 \times 11$) possible interspecies interactions. In the system of three compartments the same number of interactions must be concentrated in three compartments with only 12 ($= 4 \times 3$) possible interactions each. (The total number of possible nonzero interactions is, therefore, 36.) Thus, each compartment must have a connectance 132/36 times higher than the reticulate system's 1/12, that is, 11/36. Now, the critical connectance for a single four-species system is 1/4. Each of the three compartmented systems is, therefore, less likely to be stable than the single reticulate system. Moreover, suppose that every system with a critical connectance has on average a 0.5 chance of being stable. The chance of all three systems being stable is $(0.5)^3$, or 1/8. For all three compartments to be stable simultaneously, each must have an even lower connectance than the critical one. In short, reticulate systems are more likely to be stable than compartmented ones with identical parameters.

Now, there is no dispute that low connectance enhances the chances of a model being stable. As connectance is lowered, models will tend to become more compartmented and, because of the lowered connectance, the models are more likely to be stable. But, as already pointed out, it is possible to vary the extent to which webs are compartmented independently of connectance. A specific example is provided by Fig. 8.2, where both models have a single omnivore. In (a), the interaction is within one chain and the system is completely compartmented, while in (b) the omnivore interaction is between chains. Simulation shows that model (a) has only a 0.19 chance of being stable (Table 8.1) but, in contrast, model (b) is qualitatively stable (the $p = 3$ cycles present in (a) are removed when the omnivore feeds on a species in another food chain). This result also suggests that completely compartmented models are not the most stable arrangement. And, in this case, the result is derived from models with a biologically reasonable structure (as compared with the randomly organized models discussed so far). Is this result true for biologically reasonable models in general?

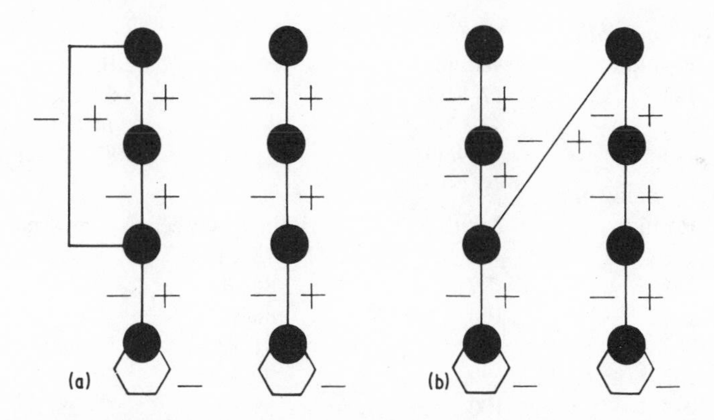

Figure 8.2 Two digraphs to illustrate differences in stability between (a) a system that is compartmented and (b) one that is not. In (a) 81 % of the models were unstable; in (b) all models are stable. Sign conventions as in Fig. 6.6.

Using the procedures discussed previously, I found that this result is true for a variety of eight-species models (Pimm, 1979b). Grouped within Table 8.1 are models that have the same number of additional interactions (those in excess of the 'basic model' of two food chains, each with four species and four trophic levels). The models are further grouped by the kind and position of these interactions. Models along each row of the table háve the same number of omnivore and non-omnivore interactions. They differ only in whether the omnivore interactions are within or between chains, that is, whether the models are more (within) or less (between) compartmented. In the table, the degree of grouping into compartments decreases from left to right. In most cases, the most compartmented models (left column) are the least stable, but the least compartmented models (right column) may be less stable than

Table 8.1 Eight-species models: percentage of unstable models. From Pimm (1979b).

Additional interactions = 0					
Only one model 0% unstable					
Additional interactions = 1					
(a) No omnivore					
				Between	
				0	
(b) 1 omnivore					
	Position	Within		Between	
	1–3	81	(Fig. 8.2(a))	0	(Fig. 8.2(b))
	2–4	91		0	
	1–4	96		0	
Additional interactions = 2					
(a) No omnivore					
				Between	
				21–51(32)*	
(b) 1 omnivore					
	Position	Within		Between	
	1–3	77–81(78)		62–81(72)	
	2–4	87–93(89)		74–87(80)	
	1–4	94–96(95)		87–95(93)	
(c) 2 omnivores					
	Position	2 within	1 within/1 between	2 between	
	1–3	96	78–81	82	
	2–4	99	86–90	92	
	1–4	100	86–94	100	
	1–3, 2–4	98	81–83	33–88	
	1–3, 1–4	99	80–82	80–96	
	2–4, 1–4	100	89–93	82–98	
Additional interactions = 4					
(a) No omnivore					
				Between	
				70–81(75)	
(b) 1 omnivore					
	Position	Within		Between	
	1–3	92		92	
	2–4	94		96	
	1–4	97		99	
(c) 2 omnivores					
	Position	2 within	1 within/1 between	2 between	
	1–3	97	93–95	97	
	2–4	99–100	97	99	
	1–4	100	99–100	100	

Table 8.1 (*continued*)

1–3, 1–4	99–100	100	98
1–4, 2–4	99–100	100	99
1–3, 2–4	99	95	97–99

(d) 4 omnivores – all greater than 99.8 % unstable

Additional interactions = 6

(a) No omnivory

			Between
			93

(b) 1 omnivore

Position	Within	← same →	Between
1–3		98	
2–4		98	
1–4		99	

2 or more omnivores: greater than 98 % unstable

* Indicates range (and mean) when various model structures are possible.

intermediate models (middle column). In short, I find May's argument that food webs should be compartmented difficult to duplicate. However, as evidenced by McNaughton's (1978) attempt to test it, it clearly appeals to the field biologist. May's suggestion is worthy of testing, with or without a theoretical basis.

8.1.2 Biological reasons

Species within a habitat might be more likely to interact with other species within that habitat rather than those outside it. Each habitat requires a set of adaptations from each of its component species, and this specialization may preclude extensive interactions between habitats. Compartments may exist for this reason and, if they do, their boundaries will correspond to the boundaries of the habitats. For similar reasons, feeding on plant detritus may preclude feeding on live plant material, and vice versa, leading to Odum's (1962, 1963) suggestion that species in the grazing and detritus food chains should be separate.

Obviously, compartments whose boundaries correspond to different habitats might arise for either dynamical or biological reasons, or both. Hence, to demonstrate that dynamic effects constrain food webs into compartments, it is necessary to examine whether species are compartmented *within* habitats. But what is a habitat? Lawton and I felt that we could not answer this completely objectively but that we could come sufficiently close to permit an evaluation of the hypotheses (Pimm & Lawton, 1980).

The definition of a habitat depends in part on the size and activity of the organism(s) in question (Elton, 1966). Southwood (1978) illustrated the problem by correlating the size of an animal with its range (Fig. 6.5). Thus, the individual grasses and shrubs in a field may represent separate habitats for herbivorous insects; but though a lark will view the same grassy field as a habitat different from a nearby wood it will ignore the changes in the abundance of the field's constituent grasses and shrubs; finally, a predatory hawk may include both the field and the wood in its territory. What constitutes a habitat will depend very much on the taxonomic group of the organism and, consequently, its trophic position.

8.2 TESTING THE HYPOTHESES: HABITATS AS COMPARTMENTS

To decide whether species interactions were more frequent within, rather than between, habitats, Lawton and I first examined insects and whether their interactions formed compartments whose boundaries corresponded to differing resources (Pimm & Lawton, 1980). Second, we examined whether major habitat divisions (forest versus prairie, for example) imposed a compartmented structure upon larger organisms. Finally, we examined interactions within and between grazing and detritus food chains.

8.2.1 Plants and their insects as compartments

Data on the herbivorous insects and their predators and parasitoids for two or more species of plant in the same local area are few. Here are three examples.

(a) The first provides quantitative as well as qualitative information about the extent of compartments. Shure (1973) studied two plants, *Raphanus raphanistrum* and *Ambrosia artemisiifolia*, that were dominant in the initial stages of an old-field succession. In different experiments, the two plant species were labelled with ^{32}P and the amount of this label found in herbivorous and predatory insects was measured over subsequent weeks.

Did the two plants and their insects behave as two separate compartments? Figure 8.3 is an interpretation of the pathways taken by the ^{32}P label, based on one experiment with *Raphanus* and three with *Ambrosia*. Thus, 54% of the label, placed initially on *Ambrosia* and recovered in herbivores, was in species that fed on both species of plant; the corresponding figure for *Raphanus* was 66%.

Shure (1973) did not indicate all the feeding preferences of the predators, so it is not certain from which of each group of herbivores (common to both plants; unique to one plant) each of the two predatory groups (common to both plants; unique to one plant) received the label. However, some of the relationships are described in his text and these suggest that predators unique to each plant species fed almost entirely on herbivores unique to each plant species. The herbivorous prey of predators common to both plants are less certain. Hence, Fig. 8.3 indicates that predators common to both plants received the label from both herbivores common to both plants and herbivores

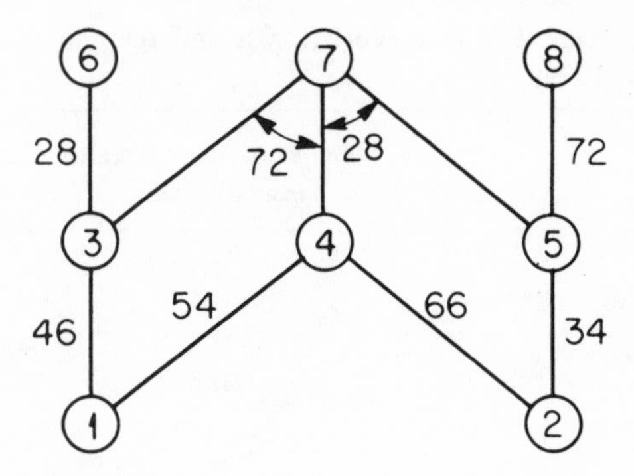

Figure 8.3 Patterns of ^{32}P tracer flow through two species of plant in an old-field ecosystem. Numbers are percentages of tracer following routes indicated; further details are given in the text. Legend: (1) *Ambrosia*, (2) *Raphanus*, (3) herbivores restricted to (1), (4) herbivores common to both species, (5) herbivores restricted to (2), (6) predators restricted to (3), (7) predators feeding on (4), and/or predators which feed on both (3) and (5), (8) predators restricted to (5). From Pimm & Lawton (1980) following original work by Shure (1973).

unique to each plant. Whichever interpretation is correct, 72% of the label placed in *Ambrosia* and subsequently recovered in predators was in species common to both plants; the corresponding figure for *Raphanus* was 28% (there is no significance in the complementarity of these percentages).

Interpreting these figures is not easy because there is no obvious null hypothesis; that is, it is not possible to know what percentages of the label should be expected in each of the groups in Fig. 8.3 under either the reticulate or the compartmented hypotheses. It could be as easily argued that the data indicate the presence of compartments as the converse. But since an average of 50% or more of the label goes to species shared by the two plants at each consumer level, our view was that the data do not support the hypothesis of two obvious compartments.

(b) The second example also involves plants and their insect herbivores. Attracted by the large number of monophagous or oligophagous herbivorous insects in certain taxa (e.g. Lawton & McNeill, 1979), we earlier suggested that food webs based on green plants might well be compartmented (Lawton & Pimm, 1978). In doing so, we overlooked the fact that each of the polyphagous species of insect associated with a particular species of plant, would feed on an idiosyncratic selection of alternative plants in the habitat. This point emerges clearly from the work of Futuyma & Gould (1979) on insect-plant associations in a deciduous forest. Table 8.2 presents the total number of insect species found on each of the eighteen common woody plants in upland forest in New York State. Only three species of plant supported monophagous herbivores and only one, the conifer, Hemlock, (*Tsuga canadensis*) had more than two

Table 8.2 Plant species in a New York forest and their insect herbivores. After Futuyma & Gould (1979).

Plant species	*Number of monophagous insect species*	*Total number of insect species*
Acer rubrum	0	27
Hamamelis virginiana	0	24
Quercus rubra	0	31
Vaccinium sp.	0	31
Quercus prinus	0	35
Quercus alba	0	29
Carya glabra and *C. ovata*	0	35
Cornus florida	1	27
Prunus serotina	2	32
Tsuga canadensis	7	18
Betula lenta	0	25
Castanea dentata	0	28
Kalmia latifolia	0	11
Fagus grandifolia	0	19
Carpinus caroliniana	0	16
Lindera benzoin	0	18
Liriodendron tulipifera	0	15
Sassafras albidum	0	13

monophages. These data do not exclude the existence of compartments based on groups of plant species such as genera (though none are obvious). They are, however, clearly incompatible with the suggestion that each plant species is the basis of a compartment. Thus, the oligophagous and, particularly, the polyphagous species may well be sufficiently important to blur any incipient compartmentation generated by the monophagous insect species on plants. When ecosystems or habitats as a whole are examined, Futuyma and Gould's work and that of others (e.g., Gibson, 1976, Hansen & Uekart, 1970, Joern, 1979 and Sheldon & Rogers, 1978) all suggest that there is a considerable overlap in the plant species used by insects. The data do not give a clear impression of compartments based on plant species.

(c) The third example is provided by the data on leaf galls and their natural enemies (mainly parasitoids and hyper-parasitoids) described by Askew (1961). All are gall-formers on the oak (*Quercus robur*). Lawton and I extracted the information for one locality – Wytham Woods, near Oxford, England, (Pimm & Lawton, 1980). These data form one of the most detailed and carefully gathered sets of data with which to test the two hypotheses. Arguably, an oak tree forms one habitat in which case the following analysis more properly belongs in Section 8.3 which considers compartments within habitats. Instead, I shall consider each kind of gall as a potentially isolated compartment (habitat) for its specific producers, lodgers and enemies. It makes no difference to the thesis which approach is adopted.

The different gall-forming species could give rise to different compartments. But examination of these data (Table 8.3) suggests that most species share many enemies, although the data also show the existence of two groups with no shared interactions (species 6, 40, and the rest of the species). Better separation might be achieved by considering certain groups of genera as compartments. For example, *Andricus* species (1–10) have a largely different set of enemies than *Cynips*, *Synergus* and *Neuropterus* species (12–26). Indeed, a number of compartments *could* be established within the data by inspection and *a posteriori* reasoning and, hence, give support for the compartment hypothesis. This is a poor procedure. The potential number of ways of organizing the interactions between these species is very large and, by chance alone, one would generate some compartments. The two compartments in Askew's data hardly seem convincing evidence when one contains only 2 species but the other contains 54. Hence, the *a posteriori* recognition of compartments is, at best, statistically risky. What it is necessary to know is whether a particular web is more compartmented than chance alone dictates. An objective statistical test with a known distribution is required, or at least one which can be approximated. Because these data span three trophic levels, it can be tested whether any set of compartments within two levels are matched by interactions between these and the third level. To do this, consider some features of Fig. 8.1.

In the system with two compartments (a), note that X_1 shares neither predator nor prey species with the species of the other compartment (X_2 and X_3). Species X_2 and X_3 share both predators and prey. The numbers of predators and prey shared are shown in Table 8.4(a). In the reticulate model (b), X_1 shares predators with X_3 and shares prey with X_2. Species X_2 and X_3 share a prey species but they do not share predators (Table 8.4 (b)). In general, there is a tendency to share prey if predators are shared and vice versa only if the system is compartmented. If the system is reticulate, then the numbers of predators shared between two species is independent of, or negatively correlated with, the number of prey shared.

There is one modification that must be made to this analysis before it can be applied to Askew's data. In the data on the oak-gall food web, there is a positive correlation between the number of predators a species suffers and the number of species of prey it utilizes. This generates a positive correlation between the number of predators and prey shared, because both these quantities depend, in part, on the number of predators and prey of each species in the pair under comparison. In such cases, the positive correlation has nothing to do with whether the system is compartmented or not. Such positive correlations between the numbers of prey and predators are unusual in food webs as a whole (Chapter 9) but, for whatever reason, one emerges very clearly from Askew's data. Fortunately, the effect of species having different numbers

Table 8.3

Prey	Predators																	
	20	21	22	23	24	25	26	27	28	29	30	31	32	33	34	35	36	37
1	2	2							1		1							
2			−1	−1	2			2		1	1	1	1					
3	2		2			−1			1					1				
4			2						−1									
5						−1								1				
6																		
7			2															
8	2		2											1				
9									−1									
10	2		2															
11			−1															
12		2	2						1				1	1				1
13		2							−1							1		1
14		2							−1		1							1
15		2												1				1
16											1							
17			−1				−1		1			1	1	1		1	1	
18	2						−1		−1			1	1	1		1	1	
19	2		2				−1		1				1					
20									1			1	1	1		1	1	
21									1			1		1		1		1
22									1					1			1	1
23									1			1	1		1		1	
24									1			1	1		1		1	
26									1							1	1	
28																1	1	
30												1						
31													1			1	1	
32														1			1	1
33																		
35																1		
36																	1	
37									1					1				
39									1					1				
42																		
43														1				
45									1									
47																		1

Row and column numbers correspond to species, as described below. Table entries are: 1, if species in column is the predator or parasite of species in row; 2, if species in column kills, but does not eat species in row; −1, if species in column eats gall in row, but does not kill it (from Askew, 1961: see text for details).

Species number	Species	Species number	Species
	Cynipidae		
1	*Andricus ostreus* (Htg.)	14	*N. albipes* (Schenck)
2	*A. kollari* (Htg.)	15	*N. tricolor* (Htg.)
3	*A. curvator* (Htg.)	16	*N. aprilinus* (Giraud)
4	*A. callidoma* (Htg.)	17	*Cynips longiventris* (Htg.)
5	*A. inflator* (Htg.)	18	*C. divisa* (Htg.)
6	*A. quercus-ramuli* (L.)	19	*C. quercus-folii* (L.)
7	*A. albopunctatus* (Schlech.)	20	*Synergus nervosus* (Htg.)
8	*A. quadrilineatus* (Htg.)	21	*S. albipes* (Htg.)
9	*A. solitarius* (Fonsc.)	22	*S. gallae-pomiformis* (Boyer de Fonsc.)
10	*A. seminationis* (Giraud)	23	*S. umbraculus* (Oliv.)
11	*Biorhiza pallida* (Oliv.)	24	*S. reinhardi* (Mayr)
12	*Neuroterus quercus-baccarum* (L.)	25	*S. evanescens* (Mayr)
13	*N. numismalis* (Fourc.)	26	*S. pallicornis* (Htg.)
		27	*Ceroptres arator* (Htg.)

38	39	40	41	42	43	44	45	46	47	48	49	50	51	52	53	54	55	56
	1				1													
	1																	
		1																
			1															
				1		1												
	1	1																
	1												1					
	1													1	1	1		
							1											
							1											
							−1											1
																	1	
								1	1									
1																		
												1						
												1						
												1						
									1	1							

Species number	Species	Species number	Species
	Chalcidoidea		
28	*Eurytoma brunniventris* (Ratz.)	42	*O. skianeuros* (Ratz.)
29	*Megastigmus stigmatizans* (F.)	43	*Syntomaspis notata* (Walk.)
30	*Mesopolobus fuscipes* (Walk.)	44	*S. apicalis* (Walk.)
31	*Torymus cingulatus* (Nees)	45	*S. cyanea* (Boh.)
32	*T. nigricornis* (Boh.)	46	*Eupelmus uroznus* (Dalman)
33	*T. auratus* (Fourc.)	47	*Caenacis divisa* (Walk.)
34	*Megastigmus dorsalis* (F.)	48	*Cecidostiba leucopeza* (Ratz.)
35	*Mesopolobus fasciiventris* (Westw.)	49	*C. semifascia* (Walk.)
36	*M. jucundus* (Walk.)	50	*Hobbya stenonota* (Ratz.)
37	*M. tibialis* (Westw.)	51	*Pediobius lysis* (Walk.)
38	*Hobbya kollari* (Askew)	52	*Ormocerus latus* (Walk.)
39	*Olynx arsames* (Walk.)	53	*O. vernalis* (Walk.)
40	*O. gallaram* (L.)	54	*Pediobius clita* (Walk.)
41	*O. euedoreschus* (Walk.)	55	*Tetrastichus aethiops* (Zett.)
		56	*Eudecatoma biguttata* (Swed.)

Table 8.4 Observed and expected numbers of prey and predators shared by three species in Fig. 8.1. Predators shared are above diagonal, prey shared below diagonal. From Pimm & Lawton (1980).

Observed (a) — A system with two compartments

Species	1	2	3
1	–	0	0
2	0	–	1
3	0	2	–

Observed (b) — A system without distinct compartments

Species	1	2	3
1	–	0	1
2	1	–	0
3	0	1	0

Expected (for both webs)

Species	1	2	3
1	–	$\frac{1}{2}$	1
2	1	–	$\frac{1}{2}$
3	1	1	–

of prey and predators can be factored out by calculating the expected number of prey (or predators) the two species should share. This requires only a knowledge of the number of species of prey (or predators) within the system. Such expected values have been calculated for the systems in Fig. 8.1 and are presented in Table 8.4. The observed values can be examined against these expected values and an evaluation made on to whether a system is compartmented or not. Such an analysis is still complicated. Note that when there are compartments the differences between the observed and expected values will be both negative or both positive for the prey and predators of the two species under comparison. The differences are independent in the reticulate model. There are four kinds of differences: two matched kinds (more prey shared than expected and more predators shared than expected; fewer prey shared than expected and fewer predators shared than expected) and two mixed kinds (more prey . . . fewer predators; and vice versa). The number of matched and mixed differences can be compared using a chi-square test. Only if the matched differences predominate is there evidence for compartments.

Finally, such an analysis of Askew's data is complicated by the fact that the complexity of the web often leads to the same species appearing in more than one list. For example, a species may appear as an intermediate species and as their predators or their prey (unlike the species in Fig. 8.1, where these are completely separate). Ignoring this complication (which probably makes little difference) shows that the difference between the observed and expected number of predators shared is independent ($\alpha > 0.05$) of the difference between the observed and expected number of prey shared. Lawton and I concluded that this system was not compartmented more than expected by chance. But we did not test whether the data differed from those expected by the reticulate hypothesis (which implies interactions more uniformly distributed than random). For this hypothesis, the degree of predators shared, given prey shared, is not obvious (Pimm & Lawton, 1980).

Comments

What kinds of information can and cannot be used to distinguish between the two hypotheses? Some studies list a taxonomically restricted set of herbivores and the plants they utilize. Cohen (1978) calls sets such as these *sink webs* and presents examples. Unfortunately, sink webs cannot be used to make inferences about compartments. Even if the species show complete monophagy, the importance of polyphagous species of different taxa is, by definition, unstudied and, therefore, unknown. Clearly, a list of plant species and *all* their herbivores is required. These lists are *source webs* and examples include the data in Table 8.2 which show that species which recognize plant species as the boundaries of their compartment are in a small minority.

If the compartment boundaries (if any) are not apparent *a priori*, then an attempt can be made to identify compartments by using the interactions between two trophic levels and then to test them on a third trophic level. This was the procedure applied to Askew's data. Comparable data are extremely scarce because few others have had the necessary time, skill and patience to collect information on large numbers of herbivores, their predators, parasitoids and hyper-parasitoids at one locality. There are numerous lists of food plants, herbivores and their natural enemies for larger geographical regions (e.g. Lawton & Schroder, 1977, Lawton & Price, 1979) but, for obvious reasons, these are not adequate to test hypotheses about food webs at particular localities within that region.

8.2.2 Vertebrates and major habitat divisions as compartments

We were able to locate four studies (Table 8.5) which described species interactions within and between major habitat divisions (Pimm & Lawton, 1980). These divisions were established *a priori* by the original authors. If the 'compartmented hypothesis' is correct, there should be fewer interactions between compartments than within compartments, relative to the number of species in each compartment. Figure 8.4 is based on the work of Summerhayes & Elton (1923) who described the major interactions within and between three habitats (terrestrial, freshwater and marine) on an arctic island. In this study, species interactions did appear to be grouped into three compartments, though evidently some species fed across the boundaries. In the four studies

Table 8.5 Observed (Q) and expected ($\hat{Q}$) numbers of interactions between two habitats (A, B) with P interactions among S species in each. From Pimm & Lawton (1980).

Habitat A	*Habitat B*	P_A	P_B	S_A	S_B	Q	$\hat{Q}$	χ^2	*Source*
Marine	Terrestrial	9	29	8	19	1	15.85	13.91**[a]	Niering (1963)
Marine	Terrestrial	21	36	26	39	11	27.36	9.78**	Koepcke & Koepcke (1952)
Aspen forest	Prairie	17	14	15	12	3	15.31	9.90**	Bird (1930)
Aspen forest	Willow forest	17	14	15	11	1	15.13	13.20**	Bird (1930)
Willow forest	Prairie	14	14	11	12	1	13.97	12.05**	Bird (1930)
Marine	Terrestrial	2	26	3	17	3	7.14	2.40 ns	Summerhayes & Elton (1923)
Marine	Freshwater	2	10	3	9	0	4.50	4.00*	Summerhayes & Elton (1923)
Terrestrial	Freshwater	26	10	17	9	5	16.30	7.83**	Summerhayes & Elton (1923)

[a] ns not significant, * significant at $\alpha < 0.05$, ** significant at $\alpha < 0.01$.

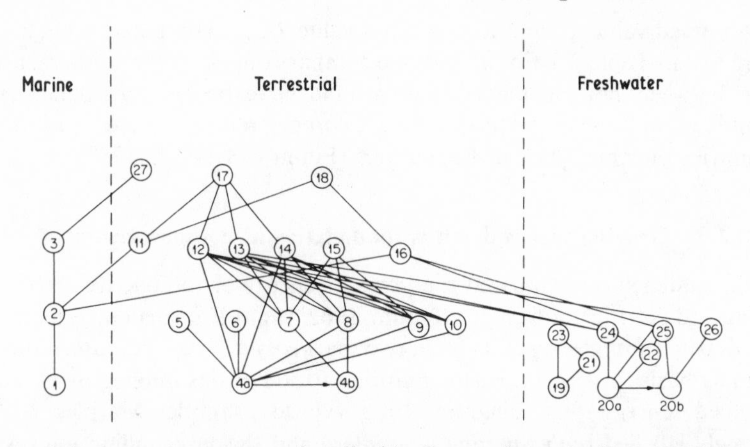

Figure 8.4 An arctic ecosystem described by Summerhayes & Elton (1923). Legend: (1) plankton, (2) marine animals, (3) seals, (4a) plants, (4b) dead plants, (5) worms, (6) geese, (7) Collembola, (8) Diptera, (9) mites, (10) Hymenoptera, (11) seabirds, (12) Snow Bunting, (13) Purple Sandpiper, (14) Ptarmigan, (15) spiders, (16) ducks and divers, (17) Arctic Fox, (18) skua and Glaucous gull, (19) planktonic algae, (20a) benthic algae (20b) decaying matter, (21) protozoa, (22) protozoa, (23) invertebrates, (24) Diptera, (25) other invertebrates, (26) *Lepidurus*, (27) Polar Bear. From Pimm & Lawton (1980).

analysed, all species were placed in only one habitat by the original authors even though some or all of their prey might come from another habitat. This suggested the following analysis.

Consider two habitats, A and B, with S_A and S_B species in each and suppose further there are P_A and P_B interactions within each habitat and Q interactions between habitats. The average number of interactions per species within compartments (habitats) is

$$(P_A + P_B)/(S_A + S_B). \tag{8.1}$$

The number of interactions between habitats should be proportional to the relative sizes of the numbers of species in each habitat, that is, proportional to

$$2\left(\frac{S_A}{S_A + S_B}\right)\left(\frac{S_B}{S_A + S_B}\right). \tag{8.2}$$

The factor 2 indicates that the interactions can go from A to B and from B to A. Since the expression in (8.1) provides a measure of the proportion of interactions to species (an underestimate because it ignores interactions between habitats, Q), this expression multiplied by (8.2) provides a conservative expected number of interactions per species between the two habitats. Thus, (8.1) multiplied by (8.2) multiplied by the number of species in the system gives the expected number of interactions between habitats, ($\hat{Q}$):

$$\hat{Q} = \frac{2(P_A + P_B)S_A S_B}{(S_A + S_B)^2}. \tag{8.3}$$

The observed value, Q, and the expected value, $\hat{Q}$, are compared using a chi-square test in Table 8.5. For all but one comparison, many fewer interactions occur between compartments than would be expected by chance. We concluded that these larger food webs are compartmented with compartment boundaries matching habitat boundaries (Pimm & Lawton, 1980).

8.2.3 Grazing versus detritus food chains as compartments

Several studies have compared the flows of energy or radioactive tracers in grazing and detritus food chains (Odum, 1962, 1963). For small invertebrates, plant detritus and green plants form two markedly different resources and so provide a basis for the compartmentation of food chains implied by Odum's 'Y-shaped' energy flow diagram. In a typical example, Marples (1966) separately labelled both *Spartina alterniflora* and the surrounding mud with ^{32}P in a salt marsh in Georgia, USA. There was virtually no overlap in the primary consumers (herbivores and detritivores) but the predators in the system, spiders, fed on both the food chains. This pattern of distinct pathways at the primary consumer level merging at the secondary consumer level (and higher) is found in a range of terrestrial and aquatic systems, as listed below:

(i) The arctic habitats of Summerhayes & Elton (1923). (In Fig. 8.4 compare the flow of energy from species 4a with 4b (live and dead plants) and species 20a with 20b (live and dead aquatic algae). In the terrestrial compartment, energy goes from 4a to species 5, 6, 8, 9, 10, 12 and 14, and from 4b to species 7 and 8; that is, only one species is shared. However, these herbivores and detritivores share exactly the same set of predators, namely, species 12, 13, 14 and 15.)

(ii) The land–sea interface of Koepcke & Koepcke (1952).

(iii) The temperate forest of Varley (1970).

(iv) The oligotrophic lake of Morgan (1972) and Blindloss, Holden, Bailey-Watts & Smith (1972).

(v) The eutrophic lake of Burgis *et al.* (1973) and Moriarty *et al.* (1973).

(vi) The river of Mann (1964, 1965) and Mann, Britton, Kowalczewski, Lack, Mathews & McDonald (1972).

In short, grazing and detritus food chains are separate compartments only at the primary producer/detritus and primary consumer levels. Above that, compartments are linked by common predators, although measuring the intensity of this linkage might be rewarding. Unfortunately, the data are not adequate to conduct an analysis comparable to that in the previous section. At best, consumers of live and dead plants appear to form ill-defined compartments linked by common predators.

8.3 TESTING THE HYPOTHESES: COMPARTMENTS WITHIN HABITATS

The previous section presented evidence for compartments in food webs based on physically distinct habitats. But such compartments were often difficult to demonstrate and sometimes they did not exist at all.

Whether 'habitat-imposed' compartments do exist within ecosystems tells nothing about whether dynamical constraints force food webs into compartments. To answer this question it is necessary to discover whether compartments exist within habitats.

The problems already encountered with Askew's (1961) data apply here *a fortiori*. It is not known where the boundaries between the possible compartments should be located, neither is the number of compartments nor the number of species per compartment known. Consequently, for any system, there might be a large range of possible compartmentations. Consider the rearrangement of model interactions between species, but with the trophic structure preserved. This produces many possible webs, among which apparent compartments might frequently be recognized. Experience suggests this to be true. To circumvent this problem, Lawton and I first defined a statistic which measured the degree to which a system was organized into compartments (Pimm & Lawton, 1980). Then we determined how this statistic was distributed in model systems under the null hypothesis that interactions are distributed at random but subject to the constraints discussed in earlier chapters. Finally, by comparing the observed values for compartmentation with those under the null hypothesis, inferences could be made as to whether real webs were more or less compartmented than chance alone dictated.

8.3.1 The compartmentation statistic

We called the statistic describing the degree to which systems are organized into compartments $\overline{C}_1$; it is only one of a large number of possible measures of compartmentation. The derivation of $\overline{C}_1$ was entirely intuitive. Its utility stemmed from its capability of correctly distinguishing the degree to which systems were grouped into compartments in a large number of test cases. It works well only when systems are compared which are identical in their number of species, connectance and a variety of other properties. This proves to be no disadvantage here, but I suggest that the statistic not be used to compare systems differing in these properties. The distribution of $\overline{C}_1$ under interesting ecological conditions is probably beyond the ability of analytical methods, but it is amenable to numerical computation.

Consider a binary matrix, $\boldsymbol{R}$, of size $n \times n$, where n is the number of species in the matrix. The elements r_{ij} are:

$r_{ii} = 1$	for all i.
$r_{ij} = r_{ji} = 1$	if i feeds on j. (Assume that all feeding links are reciprocal; that is, if i feeds on j, i influences j and j influences i.)
$r_{ij} = r_{ji} = 0$	otherwise.

Two examples are illustrated in Fig. 8.5, where it should be noted that both webs have identical numbers of species, interactions, trophic levels, top-predators and basal species. Model (a) is compartmented, model (b) is not. The matrix $\boldsymbol{R}$ (Table 8.6) shows whether a species interacts with another species. Derived from $\boldsymbol{R}$ is a matrix $\boldsymbol{S}$ (Table 8.6), which indicates the degree to which the species share predators and prey. $\boldsymbol{S}$ is analogous to a correlation matrix but uses an index of similarity more suitable for binary data than the familiar (Pearson product moment) correlation coefficient. The elements of $\boldsymbol{S}$, s_{ij}, are obtained from

$$s_{ij} = \frac{\text{The number of species with which both } i \text{ and } j \text{ interact}}{\text{The number of species with which either } i \text{ or } j \text{ interacts}}. \qquad (8.4)$$

Thus, in Fig. 8.5(a), X_2 interacts with X_1, as does X_3. Over and above this, they only interact with each other and hence $s_{23} = 1.0$. $\overline{C}_1$ is derived as the mean of the off-diagonal elements (the diagonal elements are necessarily unity):

$$\overline{C}_1 = \frac{1}{n(n-1)} \sum_{i=1}^{n} \sum_{\substack{j=1 \\ j \neq i}}^{n} s_{ij}. \qquad (8.5)$$

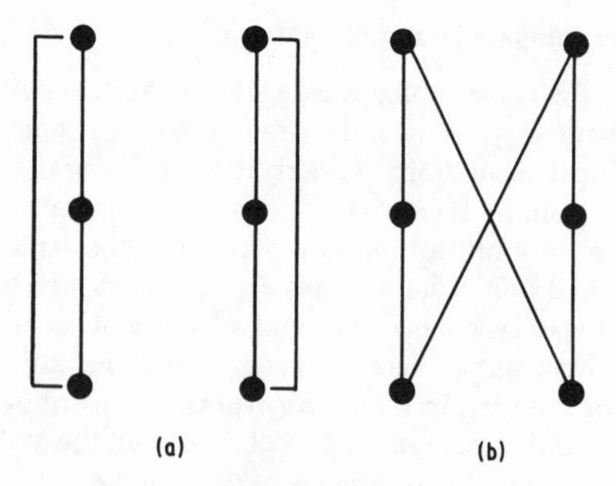

Figure 8.5 Two webs, (a) is compartmented and (b) is not, for which compartmentation statistics are derived in Table 8.4. (From Pimm & Lawton 1980.)

Table 8.6 Matrices derived from the food webs shown in Fig. 8.5 to illustrate derivation of compartmentation statistic, $\overline{C}_1$. From Pimm & Lawton (1980).

(a)

		1	2	3	4	5	6
	1	1	1	1	0	0	0
	2	1	1	1	0	0	0
R	3	1	1	1	0	0	0
	4	0	0	0	1	1	1
	5	0	0	0	1	1	1
	6	0	0	0	1	1	1

		1	2	3	4	5	6
	1	–	1.0	1.0	0.0	0.0	0.0
	2	1.0	–	1.0	0.0	0.0	0.0
S	3	1.0	1.0*	–	0.0	0.0	0.0
	4	0.0	0.0	0.0	–	1.0	1.0
	5	0.0	0.0	0.0	1.0	–	1.0
	6	0.0	0.0	0.0	1.0	1.0	–

$$\overline{C}_1 = 0.4\left(= \frac{12}{6 \times 5}\right)$$

(b)

		1	2	3	4	5	6
	1	1	1	0	0	0	1
	2	1	1	1	0	0	0
R	3	0	1	1	1	0	0
	4	0	0	1	1	1	0
	5	0	0	0	1	1	1
	6	1	0	0	0	1	1

		1	2	3	4	5	6
	1	–	0.5	0.2	0.0	0.2	0.5
	2	0.5	–	0.5	0.2	0.0	0.2
S	3	0.2	0.5	–	0.5	0.2	0.0
	4	0.0	0.2	0.5	–	0.5	0.2
	5	0.2	0.0	0.2	0.5	–	0.5
	6	0.5	0.2	0.0	0.2	0.5	–

$$\overline{C}_1 = 0.28\left(= \frac{8.4}{6 \times 5}\right)$$

* The derivation of this element is discussed as an example in the text.

This statistic is taken over the interval (0, 1). As can be seen from Table 8.6, the value of $\overline{C}_1$ is greater for the system that is more compartmented. I have checked hundreds of matrices, produced as described below, and found that the values of $\overline{C}_1$ are always ranked identically with my subjective rankings, based on examination of the actual food web structure.

Why should $\overline{C}_1$ increase with compartmentation? In the model with compartments, each species interacts or shares predators and prey with only two species, compared to four in the reticulate model. There are more zero elements in C in the model with compartments (Table 8.6), but the magnitude of the elements in the reticulate model is considerably reduced by two factors. First, the numerator (in the expression for s_{ij}) is reduced because 'i' and 'j' share fewer species of predator and prey. Second, the denominator is increased because the total number of prey and predators with which both species interact is increased.

Next, consider the distribution of the $\overline{C}_1$ statistic. The elements in the matrix $\boldsymbol{R}$ must be highly non-random for any system that is ecologically reasonable. The methods used to show this were similar (and much of the computer program identical) to those described in Appendix 6A for testing other attributes of real food webs and involved the same restrictions on web parameters and absence of loops. The webs were analysed, and their

Table 8.7 Degree of grouping into compartments. (From Pimm & Lawton, 1980.)

$\overline{C}_1^*$	*P*†	*System*‡	*Source*
0.170	0.556	prairie	Bird (1930)
0.151	0.400	willow forest	Bird (1930)
0.144	0.727	gastropods	Paine (1963)
0.264	1.00	starfish	Paine (1966)
0.337	0.402	freshwater stream	Minshall (1967)
0.175	0.015	mudflat	Milne & Dunnett (1972)
0.137	0.115	mussel bed	Milne & Dunnett (1972)
0.224	0.330	spring	Tilly (1968)
0.253	0.732	freshwater stream	Jones (1959)
0.140	0.800	lake fish	Zaret & Paine (1973)
0.135	0.333	marine algae	Jansson (1967)

* The compartmentation statistic described in the text.
† The proportion of random webs that have values of $\overline{C}_1$ that exceed that of the real web on which they are based. Details of the estimation of these values are given in this chapter and Appendix 6A.
‡ Additional parameters for these webs are given in Table 6.1.

parameters are shown in Table 8.7. For each random web, $\overline{C}_1$ was calculated and for the entire set, *P*, the proportion of random webs with $\overline{C}_1$ greater than the observed value was obtained. If the random hypothesis is correct, the mean value of *P* for all webs should be 0.5, with *P* distributed uniformly over the interval (0, 1). In other words, chance alone dictates that half of the random webs should be more compartmented than the real web and half of the random webs less compartmented. If the mean value of *P* is sufficiently small, it indicates that real webs are more compartmented than would be expected by chance. If *P* was sufficiently large, it would be concluded that the webs were more reticulate than would be expected by chance. Neither of these possibilities appears to be true. The mean of *P* is 0.49 and the probability that this result does not differ from 0.5 is considerably greater than 5%. From this test, we concluded that there was no evidence that species interactions were grouped into compartments.

8.4 FOUR COMMENTS

First, species interactions sometimes appear to be grouped into compartments, but compartments are not common. When compartments are observed, they appear for biological reasons rather than because of dynamical constraints. Encouragingly, as I pointed out earlier, a lack of compartments is in accord with predictions based on stability considerations.

Second, within habitats, food webs could be compartmented with the boundaries corresponding to the limits of most published food webs. That is, most observers of real food webs stop recording where nature provides a convenient natural compartment. Had Askew presented only one or a small number of gall webs, it would have been easy to argue that these represented a neatly defined, natural compartment. As it is, he presented numerous oak-gall webs (Askew, 1961) and it was possible to refute this challenge. Lawton and I could not refute the argument, at the next step, that the oak-gall system, *in toto*, is the compartment. But Varley's (1970) woodland web and the earlier comments about phytophagous insects, in general, make this unlikely. Those who wish to believe that webs are compartmented can always retreat to the limits of the data, but there is no clear way to establish this if published webs themselves correspond to natural compartments. Simply, more and better studies on food webs are needed, with data specifically designed to test the compartmented hypothesis. Currently, there is no evidence for compartmentation except where it is imposed by habitat boundaries. Even across habitat boundaries there are few cases where compartmentation is apparent.

Third, use of binary data (a feeding link either exists or does not exist) prevents the detection of more subtle kinds of compartments. For example, if links vary through time, compartments may come and go, although the whole web appears to be reticulate over long periods. Alternatively, compartments may exist, defined by very strong interactions and bordered by feeble ones. Paine's (1980) meticulous studies of marine intertidal food webs reveal a high variance in interaction strengths between species, and he has further speculated that 'predictable strong interactions encourage the development of modules or subsystems embedded within the community.' If this is generally so, then food webs may still be made up of relatively autonomous sub-units, despite the impression to the contrary created by binary data. There seems to be no way of establishing either the generality of Paine's proposition or the existence of compartments which vary through time, without massive research efforts. In other words, these conclusions refer to an absence of compartments in binary food web data. They do not eliminate other, more subtle kinds of compartments.

Fourth, and finally, there are some as yet rather uncertain implications for species deletion stability in these results. Recall that species deletion stability is defined as the probability that no other species will become extinct following the removal of a species from the system (Chapter 3). Species deletion stability is generally low (approximately 0.3) and details of food web design generally have little effect on its magnitude (Chapter 4). Experimental removal of species from systems often results in other species extinctions which cause, in turn, even more species extinctions (Table 5.4). How many species 'ought' to become extinct under each hypothesis, I cannot say. But, other things being equal, ecosystem disturbances can propagate further under the reticulate hypothesis than under the compartmented hypothesis. Under the compartmented hypothesis, species losses must be limited to one compartment because, by definition, there are no interactions between compartments.

8.5 SUMMARY

In general, randomly constructed food webs are less likely to be stable the more species they contain, the more interactions between species and the greater the intensity of these interactions. In addition, May (1972, 1973b) has argued that, given these three parameters, model food webs have a better chance of being stable when interactions are arranged into compartments. Compartments exist if the interactions within a web are grouped so that species interact strongly only with species within their own compartment and interact little, if at all, with species outside it.

May's theoretical arguments are hard to duplicate and certainly do not hold for food web models that exclude many of the biological absurdities of random models. For biologically reasonable models, the most compartmented models are usually the least stable. None of this denies the biological appeal of May's hypothesis.

There are biological as well as dynamical reasons why food webs should be compartmented. Species within habitats are more likely to interact with each other because of the range of adaptations which enhance a species' success within the habitat but which preclude it from feeding successfully outside it. Thus, only if food webs are compartmented within habitats is there evidence for May's assertion that dynamical constraints cause compartments.

What constitutes a habitat depends very much on the size of the animal. I start by considering whether interactions between insects are compartmented, according to the food plant of the insect herbivores. The answer seems to be 'no'. In contrast, interactions between vertebrates do seem to be compartmented with the compartments corresponding to such major habitat divisions as 'forest' and 'prairie'.

Finally, I examine whether food webs are compartmented within habitats, which is the only way to decide whether compartments do, or do not, exist for dynamical reasons. I derive an index of compartmentation and show that real food webs do not differ, in this statistic, from webs created randomly within biologically reasonable constraints. Food webs are not, therefore, compartmented within habitats. This reaffirms the usefulness of the predictions of food web design made from considerations of the dynamics of biologically reasonable models. Yet again, what is predicted from dynamical constraints on models is observed in nature.

The conclusion that food webs are not commonly compartmented rests on binary data, that is, whether two species interact or not. The data do not permit an evaluation of Paine's (1980) idea that, when the strength of species interactions are taken into account, only species belonging to the same compartment will interact strongly.

9 Descriptive statistics

In this chapter I present three more attributes of food webs. First, I examine the ratio of the number of species of predators to the number of species of prey in a web. Next, I ask: what is the correlation between the number of predatory species (at trophic level $n + 1$) which a species (at trophic level n) suffers and the number of species of prey (at level $n - 1$) which the species exploits? Finally, I review Cohen's (1978) monograph on the dimensionality of niche overlaps.

These three features have an attribute in common and one that distinguishes them from the other features discussed so far. In previous chapters I have usually presented the theory first and then examined real food webs to see if they confirmed the theory. This has been the correct perspective historically with one exception, namely food chain lengths, where theory and observation have proceeded more or less simultaneously. The attribute common to the features I discuss in this chapter is that there is little theory to accompany the observations. In particular, though the features were uncovered on the basis of some insight or hunch, the theoretical bases for the observations are relatively less compelling than the observations themselves. Thus, in constrast to the development in earlier chapters where theoretical results demanded observations to test them, the emphasis in this chapter shifts to observations that require at least an expansion of their theoretical bases in order to be fully explained. Of course, the two approaches – theory suggesting observation, observations suggesting more theory – are but arbitrary stages in the constant cycling of theory, observation and back to theory that is the central feature of the scientific process. The approach of using observations to suggest theory has the obvious advantage that it picks up the most obvious attributes of food webs. Starting with theory may yield predictions of web features that, though present in the real world, differ only slightly from random webs. But there are special problems when we are influenced largely by observations. The patterns found may be statistical artifacts – consequences of 'staring too hard' at random distributions. Apparent patterns are seen even in the random scattering of stars in the night sky (for example, for the constellation Ursa Major, every Englishman sees a plough and every American a water dipper). Additionally, the biases of the observer and the way he or she presents the data may lead to unexpected patterns. Of course, such errors can also affect the tests of theory; but the problem is more acute when the data are being exhaustively screened for possible patterns.

The consequence of all this is that I do not give much theoretical explanation after the description of each food web pattern. Rather, I concentrate on evaluations of whether the proposed patterns are artifacts. Only when they can be certain that the proposed patterns are real will theoreticians try to explain them.

9.1 PREDATOR–PREY RATIOS

9.1.1 Cohen's result

Cohen (1977, 1978) has argued that the ratio of the number of kinds of prey to the number of kinds of predators in community webs seems to be a constant near 3:4. In particular, the ratio is less than unity. Cohen notes that this situation can pose some theoretical difficulties. As I discussed in Chapter 2, with a two trophic level system more species of predator than prey (e.g. Fig. 2.4) leads to a singular matrix; an equilibrium is not possible. With more than two trophic levels, stable webs with predators in excess of prey are quite easy to arrange. The example shown in Fig. 9.1 is, moreover, qualitatively stable yet has three predators (X_2, X_3 and X_4) but only two prey (X_1 and X_3). None the less, the ratio of predators to prey is an interesting property in its own right, especially since it appears so constant. First, I shall discuss how Cohen obtained his result, then I shall present my objections to it.

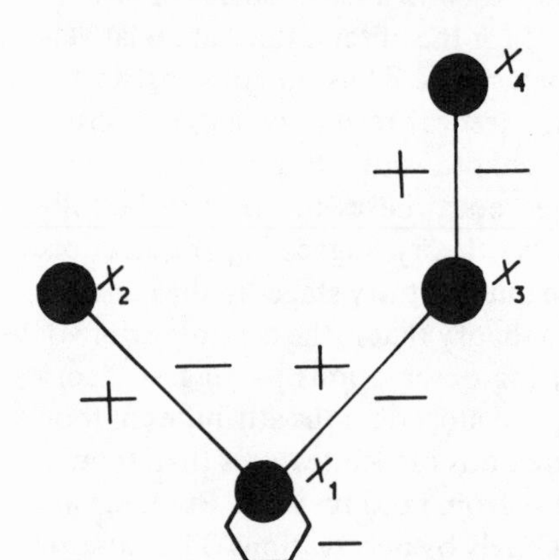

Figure 9.1 A digraph of a qualitatively stable food web with more species of predators than prey. Sign conventions as in Fig. 6.6

First, Cohen addressed the components of his webs. As I have noted earlier, these components are not individual species but kinds of organisms. A 'kind of organism' may be a stage in the life cycle or a size class within a single species, or, more commonly, a collection of functionally or taxonomically related species. Thus, the numbers in his analyses refer only to these ecologically defined kinds of organisms and not necessarily to any conventional taxonomic unit. As I have done in previous chapters, Cohen defined a predator as a kind of organism which consumed at least one kind of organism included in the food web. A prey was defined as a kind of organism which was consumed by at least one kind of organism in the food web. Clearly, some kinds of organisms were both predators and prey.

Cohen characterized food webs as belonging to one of three types: community, sink, and source webs. *Community* food webs describe organisms in a habitat, without reference to the feeding relationships among them. *Sink* food webs describe all the prey taken by a set of one or more selected predators plus all the prey taken by the prey of those predators and so on. *Source* webs describe all the predators on a set of one or more selected prey organisms plus all the predators on those predators and so on. (Sink and source webs were excluded from Cohen's study. Cohen found only one source web.) Sink webs must clearly overestimate the prey–predator ratio, for a sink web includes all the prey of the chosen predators without necessarily including all the other predators of those prey. Only community webs do not selectively omit predators. In sink webs, the prey–predator ratio averages approximately 3.5:1 (Cohen, 1978, Fig. 10).

Cohen's data on community webs are presented in Table 9.1, in which I have made some changes and included only the simplest (and usually the most certain) versions of each web. I have omitted two webs: (i) Bird's (1930) aspen–parkland web, which is a simple composite of two other webs he presents; and (ii) Teal's (1962) Salt-marsh web, which is little more than a restatement of an ecosystem as two food chains based on plants and detritus

Table 9.1 Ratio of prey to predators in community webs. (From Cohen, 1978.)

*Web number**	*Reference*	*Type of system*	*Number of prey*	*Number of predators*	*Ratio*
1.1	Bird (1930)	prairie	10	14	0.71
1.2	Bird (1930)	willow forest	8	8	1.00
1.3	Bird (1930)	aspen forest	19	22	0.86
2.0	Clarke, Flechsig & Grigg (1967)	marine, sandy bottom	10	11	0.91
7.0	Koepcke & Koepcke (1952)	sandy beach	41	50	0.82
11.0	Niering (1963)	coral atoll	18	19	0.95
15.0	Summerhayes & Elton (1923)	Arctic island	23	24	0.96
18.0	Minshall (1967)	stream	8	11	0.73
25.0	Harrison (1962)	rain forest	5	8	0.63
28.1	Fryer (1957)	rocky shore of lake	14	28	0.50
28.2	Fryer (1957)	sandy shore of lake	18	34	0.53
28.3	Fryer (1957)	creek	15	28	0.54

* From Cohen (1978).

with inputs of detritus coming from the dead tissues of most of the components of the web. (His categories are too lumped and too few to be usable.) The 12 webs left from Cohen's 14 have a mean prey–predator ratio of 0.76, which is significantly less than unity ($t = 4.38$, $\alpha < 0.01$). Statistically, there is an excess of predators in these webs. Cohen objected to calculating a statistical significance on the grounds that different webs, collected by the same author on similar systems, are unlikely to be truly statistically independent. This seems a distinct possibility when it is considered that all of Fryer's (1957) webs have similar and usually low prey–predator ratios. The mean prey–predator ratio based on the mean value for each of the eight studies does not alter the conclusion that predators are in the excess.

9.1.2 Is this result an artifact?

Despite the statistics, I have my suspicions about this result. The reason is the widespread antipathy among ecologists towards plant and invertebrate taxonomy. The 'kinds of organisms' in food webs differ greatly in the degree of taxonomic lumping and in a systematic way. Frequently, the lowest trophic levels include 'plants' or 'phytoplankton', the second 'insects' or 'invertebrates' or 'zooplankton', yet the highest levels include the full scientific names of birds, fish or mammals. The description of a sandy bottom community by Clarke, Flechsig & Grigg (1967) is quite typical. The kinds of organisms at the base of the web are: (1) small fishes and invertebrates; (2) ophiuroids (brittle stars); (3) polychaetes; (4) benthic crustacea; (5) hypoplanktonic crustacea; and (6) zooplankton. The kinds of organisms at the top of the web are all named to species (12 in all) with one exception that involves five species of perch. Fryer's (1957) webs have the same properties: the kinds of organisms at the base of the web are phytoplankton, insects or algae, and the species at the top of the web are all named specifically. Even when the plants are named specifically, there is a tendency to name only a few of those that must occur in the community. In contrast, a large proportion of the vertebrates present usually appear in the web. The effects of this preoccupation with vertebrates seem clear. There must be many more species missing from the base of the web than from the top. Webs that suffer this bias must have prey–predator ratios that are too low.

Simply, there is nothing wrong with Cohen's analysis, and the question he poses is certainly an interesting one. The data, however, are inadequate to answer this question. The solution is to obtain better data, but currently detailed studies of community webs are few. Askew's (1961) study of galls (Table 8.3) is as complete a description of a web as I can find. In every case he names the species for the 38 prey and the 37 predators. Thus, in the one case I can find where the species are not lumped into 'kinds of organisms', the excess of predators is not apparent.

9.2 THE NUMBER OF SPECIES OF PREY THAT A SPECIES EXPLOITS AND THE NUMBER OF SPECIES OF PREDATOR IT SUFFERS

What should be the relationship between the number of species of prey that a species exploits and the number of predatory species that this species suffers? My initial guess to this question was that the more species of prey a species exploits, then the more predators it should be able to support. Figure 9.2 shows such a web, where species X_1 has four prey and two predators, and species X_2 has two prey and only one predator. This idea can be developed further by considering a system of three trophic levels with a species, X_q, having n prey at the trophic level beneath it and m predators at the trophic level above it. Under Lotka–Volterra dynamics, X_q can invade this system at low densities if $\dot{X}_q > 0$. This requires that

$$\sum_{j=1}^{n} a_{qj} X_j^* > - \sum_{i=1}^{m} a_{qi} X_i^* + b_q. \tag{9.1}$$

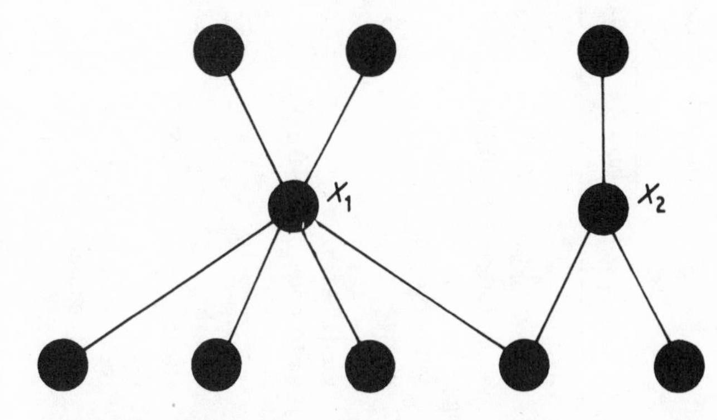

Figure 9.2 A food web where X_1 has both more species of predators and prey than X_2.

The minus sign on the right hand-side occurs because the predatory effects, a_{qi}, are negative. This is a necessary condition for X_q to belong to the community. It is not sufficient because, having invaded, X_q may cause the previously stable equilibrium of $n + m$ species to become unstable. In biological terms, X_q may invade and cause extinctions. None the less, expression (9.1) should give some idea of what conditions should be met.

If the X^* and the a_{ij} parameters are not functions of n and m (an assumption to which I shall return), inequality (9.1) implies that there should be a positive correlation between the number of prey and the number of predators. In other words, if m is large then so is n. Surprisingly, the data show the opposite to be true: if m is large then n is small, and vice versa.

9.2.1 The data, analyses, and results

Shown in Table 9.2 is my analysis of 25 food webs. In addition to the omissions discussed in Appendix 6A, I have also excluded webs that were too simple to

Table 9.2 The correlation of the number of species of predator and the number of species of prey interacting with species in various webs.

Web number*	Source	Type of web	Correlation coefficient (R^*) Versions†			P_i, Probability $R > R^*$‡		
			None	*Simple*	*Complex*	*None*	*Simple*	*Complex*
C1.1	Bird (1930)	prairie		−0.191	−0.200		0.259	L
C1.2		willow forest		−0.471	−0.845		0.867	L
C1.3		aspen forest	−0.330			L		
C2	Clarke Flechsig & Grigg (1967)	sandy bottom		S§	−0.123			L
C5	Hardy (1924)	marine		S	+0.214			L
C7	Koepcke & Koepcke (1952)	sandy beach		−0.024	+0.007		L	L
C12	Paine (1963)	gastropods	−0.681			0.926		
C15	Summerhayes & Elton (1923)	Arctic		−0.239	−0.279		L	L
C16.2	Paine (1966)	starfish		−0.811	−0.486		0.984	0.725
C18	Minshall (1967)	stream	−0.505			0.602		
C19	Valiela (1969)	dung		S	−0.752			L
C28.1	Fryer (1957)	rocky shore	+0.009			L		
C28.2		sandy shore	−0.268			L		
C28.3		stream	+0.154			L		

P1	Hurlbert, Mulla & Willson (1972)	pond	−0.750			0.820		
P2	Tilly (1968)	spring	−0.320			0.625		
P3	Milne & Dunnett (1972)	mudflat	−0.070			0.515		
P4	Jones (1959)	stream		+0.923	+0.536		0.021	0.009
P5	Zaret & Paine (1973)	lake	+0.463			0.156		
P6	Woodwell (1967)	estuary	+0.510			L		
P7	Root (1973)	collards	−0.378			0.818		
P8	Askew (1961)M §	oak galls	+0.544			L		
P9	Force (1974)	*Baccharis*		−0.515	−0.979	ND		
P10	Mayse & Price (1978)	soybeans		−0.493	−0.698		0.750	L
P11	Readshaw (1971)	orchards	0.000	−0.415		ND		

* Web numbers marked C are those of Cohen (1978); those marked P are additional webs.

† Simple versions exclude possible interactions, complex versions include them; in C16.2 and P4 interactions which account, in total, for less than 5% of the predator's diet are excluded from the simple versions.

§ Additional notes: L, web is too large (or complex) to perform the analysis; S, web is too simple in that there may be less than four species on which to perform analyses and/or all species have fixed number of prey species; M, modified as described in Pimm & Lawton (1980); ND, web contains loops of the kind A eats B eats A, when trophic levels are not defined.

‡ The proportion of random webs (with constraints as described in Appendix 6A) that have a correlation coefficient that exceeds that of the real web.

generate a correlation coefficient. These were webs that had less than four intermediate species (species that were both predators and prey in the web) or webs in which the intermediate species all had only one prey or one predator species. Webs excluded on these grounds were those of Milne & Dunnet (1972), Jansson (1967), Birkeland (1974) and Rejmanek & Stary (1979). The correlation coefficients between the number of predators and the number of prey of the intermediate species are shown in Table 9.2. I analysed nearly half the webs in two versions: a simple one, which ignores interactions labelled by the authors as probable; and a complex one which includes these interactions. Finally, some webs included quantitative information and I derived simple versions of these by excluding interactions that accounted for less than 5% of the predator's diet in total (see Appendix 6A).

There are two sets of data, namely webs with no versions plus simple versions and webs with no versions plus complex versions. Their mean correlation coefficients are shown in Table 9.3. The first set has an average correlation that is negative and very close to being significant ($\alpha = 0.06$, a two-tailed 't' test) and the second set has a significant negative correlation ($\alpha = 0.01$).

These tests assume a null hypothesis of a zero correlation, which is conservative. The more that is known about a species, the more predators and prey it will appear to have. An example of this is the study by Koepcke & Koepcke (1952) of a web involving the animals on and around a sandy beach. A crab (*Ocypode quadichaudii*) had more prey than any other species in the system (13) and almost the maximum number of predators (11) (Cohen, 1978, p. 140). While this may reflect reality, it is more probable that both these high

Table 9.3 Summary statistics for Table 9.2.

Statistic	*Correlation coefficients*	
	Case	
	None and simple	*None and complex*
Number of observations	22	25
Mean correlation	−0.156	−0.225
Standard error	0.086	0.072
	Probabilities, P_i	
	None and simple	*None and complex*
Number of observations	12	9
Degrees of freedom	24	18
$\chi^2 = \left(\sum_{i=1}^{n} -2\ln(P_i)\right)$	19.40	18.01
Significance (α)	$\simeq 75\%$	$\simeq 55\%$

numbers reflect the authors' greater interest in the crab than in the other species. A consequence of this is a positive correlation of interest (and effort) with both the numbers of prey and predators recorded for each species. These two positive correlations lead to an expectation of a positive correlation between the numbers of species of prey and predator for each species, even if no biological factors affect this phenomenon. Further examples of this positive correlation come from two webs which provide quantitative data: those of Jones (1959) and Askew (1961). Interestingly, these two webs have the largest positive correlation coefficients in the data. And in both cases there is a strong correlation between the numbers of individuals of species sampled and both the number of predators and prey recorded for those species. In short, a positive correlation between a species' predators and prey seems a likely outcome if there is an uneven sampling effort. Thus, the observed average negative correlation is even more surprising.

9.2.2 Is this result an artifact?

When I first considered this problem, the arguments presented above made me feel that the answer to this question must be 'no'. In particular, the biases due to differential lumping, discussed in section 9.1.2, seemed to have no effect. With a food web such as that of Fig. 9.2, it hardly matters if the members of the first trophic level are plant orders and the members of the third trophic level are individual species. These errors will affect both the intermediate species equally. Unfortunately, this is not the way to examine the problem. On average, there is a negative, and not a positive, correlation between prey and predators, and so a web that yields such a negative correlation should be considered. Four species arranged as shown in Fig. 9.3(b) provide an example. The species at the second trophic level has one prey and two predators, while the species at the third trophic level has one predator and two prey.

Now, I wish to consider why differential lumping of species into groups may yield a negative correlation which is an artifact. But first, let me consider the differences between webs that yield positive (Fig. 9.2) and negative correlations (Fig. 9.3(b)). The positive correlation expected in the introduction to this topic (and Fig. 9.2) and the negative correlation in this example (Fig. 9.3(b)) are not strictly alternatives. The first expects a positive correlation for species *along* one trophic level, the second a negative correlation for species *across* different trophic levels. Indeed, it may be possible to find a positive correlation along trophic levels and a negative correlation between trophic levels in the same web. Which web shape predominates (a 'short, fat' web, with many species at few trophic levels, or a 'tall, skinny' one, with the reverse) will determine whether the overall correlation is positive or negative. And which shape predominates may be a function of the observer's bias, the real shape of the web, or both. The observer's biases are often dramatic. Some webs are entirely of species at two trophic levels (and cannot be used in this instance), while others are little more

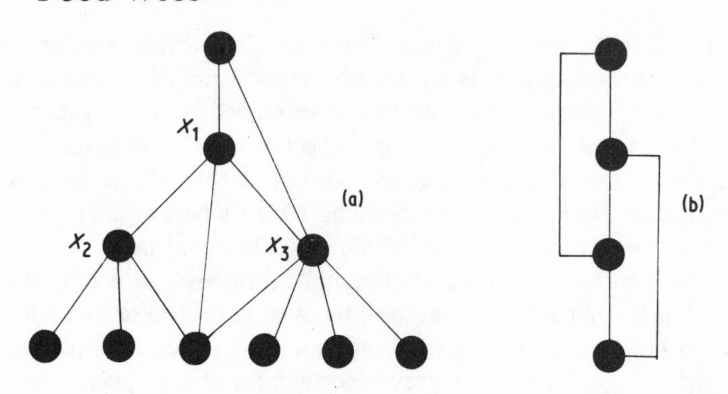

Figure 9.3 (a) A food web where a positive correlation exists between the number of species of prey the X_i exploit and the number of species of predators the X_i suffer. (b) Web (a) redrawn with all the basal species grouped into one 'kind of organism' and species X_2 and X_3 also grouped. In (b) there is a negative correlation between the numbers of predators and prey.

than four trophic levels with one species per level. Few webs are 'well proportioned' with many species at each of several trophic levels.

In the data, the negative correlation predominates. Can it also be a consequence of the same vertebrate-centered view of ecosystems, discussed earlier? As pointed out, kinds of organisms at lower trophic levels will be more lumped than those at higher levels, and this selective reduction can reverse a correlation. Figure 9.3(a) is an example of the 'real' web and that of 9.3(b) is the lumped version derived from it. In the real web, there are three species with both predators and prey. These species, with their numbers of species of predators and prey in parentheses, are: X_1 (1, 3), X_2 (1, 3) and X_3 (2, 4). The correlation between prey and predators is positive. In the lumped version, six plants are now one kind of organism and species X_2 and X_3 are also lumped. A negative correlation results.

In short, a negative correlation is expected when webs are 'tall and skinny', and such webs may be the product of the extensive grouping of species into kinds of organisms that is prevalent at lower trophic levels.

The null hypothesis

Is there any evidence that the negative correlations observed are simply functions of the shape of the food web as it is presented? If so, then because the shape of the web reflects, in part, the observer's bias, the overall negative correlation may be an artifact. Given the shape of the web, are the negative correlations observed usual? To answer this question, I used the same recipe described in previous chapters. I created random webs that had the same number of top-predators, intermediate species, basal species, interactions and trophic levels as the real web. From this sample of random webs, I calculated the correlation coefficients, R, and the proportions of R that were smaller, equal to, or exceeded those of the comparable real correlation coefficients, R^*. Table 9.2 shows the proportions that exceeded R^*, which were designated P_i. If these proportions were generally small (on average, less than 0.5), I could conclude that, even though the actual correlation coefficients are usually negative, they were more negative than one would expect by chance given the general structure of the web. But this is not the case, for the webs are not different from what would be expected by chance (Table 9.3 also shows the results of the Fisher test on proportions discussed on page 129). Overall, these results suggest that real webs are not unusual in having negative correlations of predators and prey: what is observed is what would be expected from the shapes of the webs. Thus, the shape of a food web is, on average, sufficient to explain statistically the observed negative correlation between the species' predators and the species' prey. Because the shape of the food web may reflect observer bias, I conclude that the negative correlation may well be an artifact too.

Yet, tall, skinny webs with genuine negative correlations are ecologically plausible. Web shape is a reflection of the observer, but equally certainly some webs could have shapes that would genuinely lead to negative correlations. If there are negative and genuine correlations, what is their significance?

9.2.3 The consequences of a negative correlation

Consider inequality (9.1) with the knowledge that n and m are negatively correlated. One or both of two possibilities must hold:

(i) As the number of predators m increases, although the number of prey species n decreases, the benefit the species X_q obtains from these prey increases. That is, on average the product $a_{qj}X_j^*$ increases. (Recall that this product is the *per capita* effect of the prey upon the species multiplied by the prey's equilibrium density.)

(ii) As the number of predators increases, so their average effect decreases, that is, the product of the *per capita* effect of the predator on the species multiplied by the predator's equilibrium density decreases.

Stated this way, both of these trade-offs seem reasonable. Indeed, they are little more than a restatement of the idea that there is a trade-off between specialization and generalization. Thus, X_q might either exploit a few prey efficiently or many prey inefficiently – it might be a 'Jack of All Trades' or a 'Master of One'. Explicit tests of this familiar idea are few, but it is at least possible. The possibility leads to the suggestion that, even for species along a trophic level, there may be a negative correlation and not the positive one suggested above.

What about the evidence for (i) and/or (ii) across trophic levels? I have suggested that the negative correlation might arise from species high in the web having few predators and many prey, and species low in the web having the reverse. Possibility (i) implies that species low in the web should either be more efficient at exploiting their prey or should exploit prey with higher equilibrium densities than those species high in the food web, or both. The latter alternative is usually true and is embodied by Elton's familiar 'pyramid of numbers' concept. The first alternative is also probable: I discussed in Chapter 6 how endotherms, vertebrate ectotherms and invertebrate ectotherms have increasing ecological efficiencies. Generally, endotherms are at a higher trophic level than vertebrate ectotherms which are at a higher level than invertebrates. In short, the negative correlation between a species' predators and prey does seem reasonable. If it exists, it might be a consequence of the patterns of ecological energetics discussed earlier.

If there really is a negative correlation, then there would be interesting implications for the third hypothesis on the length of food chains (discussed in Chapter 6). Recall that this hypothesis states that food chains may be short because species feed as low in the food chains as possible in order to obtain the most energy. Of course, if species low in the food chain have fewer prey and suffer more predators, then feeding low in the food chain will not be nearly as attractive as this hypothesis suggests!

9.3 INTERVAL AND NON-INTERVAL FOOD WEBS

If the patterns of prey used by predators in a web can be expressed as possibly overlapping intervals along a line, then the overlaps are deemed interval. For simplicity, the webs to which such predators belong are called 'interval webs'. If this property does not hold, the overlaps (and the webs) are considered non-interval (Cohen, 1978). For example, consider the webs in Fig. 9.4. In (a), X_1

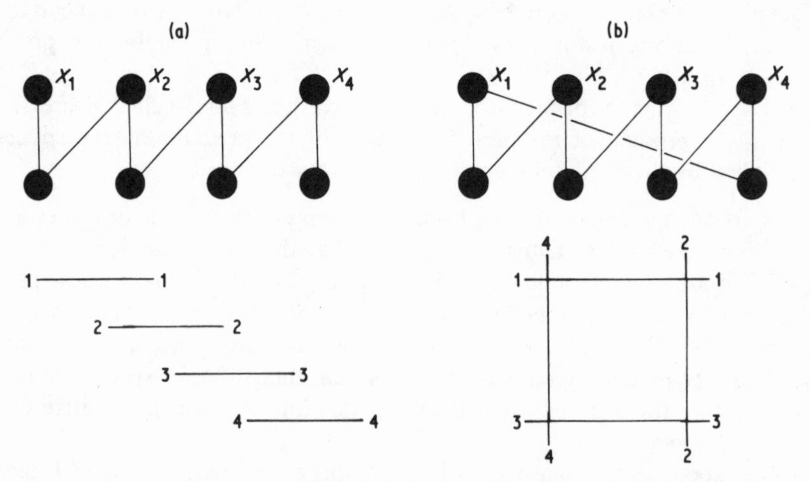

Figure 9.4 (a) An interval food web and (b) a non-interval food web. Lines below the figures show the patterns of overlap among the four predatory species.

shares prey with X_2, X_2 with X_3, and X_3 with X_4. These overlaps can be expressed by drawing segments of a line representing each predator and showing the same patterns of overlap as the predators they represent. An example of this is given below the web. Now, in (b), in addition to these overlaps, species X_1 also overlaps with X_4. Such a pattern cannot be expressed by segments on a line, though as I show below the figure, the pattern can be expressed by overlapping lines in a plane.

The non-interval pattern requires at least four predators, as three or less will always result in interval webs. Even with four predators, however, the pattern is fragile. Additional overlaps (say, between X_1 and X_3, X_2 and X_4, or both) will make the web interval again, as will fewer overlaps. Some patterns with four overlaps are also interval: for instance, were the fourth overlap to be any other than that between X_1 and X_4 then the web would be interval again. Simply, non-interval webs are not that likely when there are few predators, though as I shall show, they are likely when there are many predators. Cohen (1978) concludes that interval webs are *so* common in the real world that they are commoner than would be expected by chance. This means that the feeding relationships between predators can be represented in one dimension (not yet identified) far more often than would be expected by chance. I shall now discuss how Cohen reaches this conclusion.

9.3.1 Are food webs interval more often than would be expected by chance?

(a) Null hypotheses

Cohen's data are presented in Table 9.4. Many of the studies in the table are webs I have discussed before, while others are webs with only two trophic levels which, though adequate for an examination of this topic, were not suitable for calculating food chain lengths or the patterns of omnivory. Some webs come in several versions which reflect whether interactions considered 'possible' by their authors were included or not. Of the 28 webs in Table 9.4, all but seven (eight in some versions) are interval. Another three webs, discussed by Cohen, have less than four predators and are inevitably interval. Are these more interval webs than would be expected by chance?

The problem is the now familiar one of producing an expected distribution of a food web attribute under a null hypothesis. Cohen had seven such hypotheses differing, more or less, from the ones I have used. Cohen's hypotheses reflected an interest in patterns of niche overlap, rather than my concern about lengths of food chains and omnivory. His models involved the m predators in the web and the n prey, as well as the number of interactions n_{int}, as did mine. For each of his models, Cohen calculated a number E (for 'edges') that is the number of pairs of species that share one or more prey, that is, the number of overlaps. He then computed the observed and expected number of

Table 9.4 Interval and non-interval webs and the chance of interval webs in a random sample. (After Cohen, 1978.)

*Web number**	*Proportion of interval webs in model webs†*	*Number of predators in real web*	*Is real web interval?*	*Type of system*
1.1	0.5–0.15‡	14	yes	prairie
1.2	0.70–0.77	8	yes	willow forest
1.3	0.03	22	yes	aspen forest
1.4	0.00	25	no	1.1 and 1.3
2.0	0.00–0.19	11–18	yes	marine, sandy bottom
4.0	1.00	4	yes	salamanders
5.0	0.00–0.97	6–18	yes	herring and its food
7.0	0.00	50–58	no	sandy beach
8.1	0.22	9	yes	*Conus*: marine benches
8.2	0.00	13	no	*Conus*: reefs
8.3	0.00	13	no	8.1 and 8.2
10.0	0.99–1.00	6	yes	starfish
11.0	0.00	19	yes/no ‡	coral atoll
12.0	0.72	8	yes	gastropods
13.0	0.74	9	yes	pine trees and insects
15.0	0.00	24–27	no	Arctic island
16.2	0.79	7	yes	starfish
18.0	0.32	11	yes	stream
19.0	0.85–1.00	6	yes	dung
20.0	0.96	6	yes	*Chaetognatha*
23.0	1.00	4	yes	lake triclads
24.0	0.98	5	yes	salt marsh
25.0	0.81–0.83	8	yes	rain forest
27.0	0.99	11	yes	river
28.1	0.00	28	yes	lake: rocky shore
28.2	0.00	34	no	lake: sandy shore
28.3	0.00	28	no	stream
29.0	0.98	5	yes	marine fish

* After Cohen (1978).
† Proportions are using Cohen's model six and come from a sample of 100.
‡ Some webs come in several versions depending on whether probable or possible interactions are included or not. Details are given by Cohen.

overlaps to show that only one of the models provided a good quantitative description of the observed number of overlaps. This was model six. The other models were systematically biased in the number of overlaps they produced.

Model six involved a binomial distribution of interactions within the web. Biologically, the model assumed that every predator in the web had a constant and independent probability of preying on every prey within the web. Each of the $n \times m$ possible combinations had a probability of being an interaction of

$n_{int}/(n \times m)$. Another of Cohen's models (model five) was similar and also a reasonable predictor of the number of overlaps. The results of this model are qualitatively similar to the ones based on model six that I shall discuss presently. Cohen's model five, however, is very similar to the models I have used in that the number of interactions was fixed and randomly distributed among the $n \times m$ possible interactions. In Cohen's models, predators can be without prey and prey can be without predators; in mine they cannot. But, from my experience, these possible outcomes are infrequent. My models start with the same randomizations as Cohen's and screen out these possibilities. Very few of my randomizations fail these extra tests. In addition, I set the randomizations to avoid loops, whereas Cohen does not. Nor does Cohen restrict the number of trophic levels in his randomizations. However, although Cohen's model five is different from my model in some respects, both models have many features in common and I doubt that the differences in our approaches would lead to different conclusions.

(b) Results

For each version of each food web, Cohen produced 100 artificial webs according to the recipes of each of his six models. Each web was then tested to see if it was interval or not using a computer algorithm which Cohen describes in his Chapter 3. Table 9.4 shows the proportion of model webs that were found to be interval using model six. None of the model webs based on the eight webs that were non-interval was ever found to be interval: the proportions of interval webs were all 0.0. The non-interval condition was thus the most likely condition for a web of their given size and complexity. But, model webs based on the real webs that were interval had varying proportions that were interval. Cohen showed that these proportions were, on average, so small that it was improbable that so many interval webs could occur by chance. Real webs had a statistical excess of patterns of interval overlaps. This conclusion was unaltered by the choice of food web versions or by the selection of the various models used to produce the randomized webs.

9.3.2 Explanations and consequences

(a) Which webs are interval?

Cohen goes to considerable lengths to show that his results are not trivial and I shall not repeat his arguments here. The effect of differential lumping of species at different trophic levels cannot affect his results in any simple way. As I discussed earlier, too many interactions can prevent a non-interval pattern but, then, so can too few. Moreover, such effects occur in both the real web and the model webs because the latter have the same number of kinds of organism which are present in the real web. The biases are likely to appear equally in both the observed and expected sides of the equation.

There is, however, an interesting relationship between the proportion of model webs that are interval and the number of kinds of predators in the web (Fig. 9.5). Below six predators, nearly all the models are interval, as they must be, of course, for below four predators. Between six and 12 predators, there is a sharp drop, and above 12 nearly all the models are non-interval. Despite the variety of webs used to produce the relationship shown in Fig. 9.5, it shows little scatter. The relationship has a number of consequences.

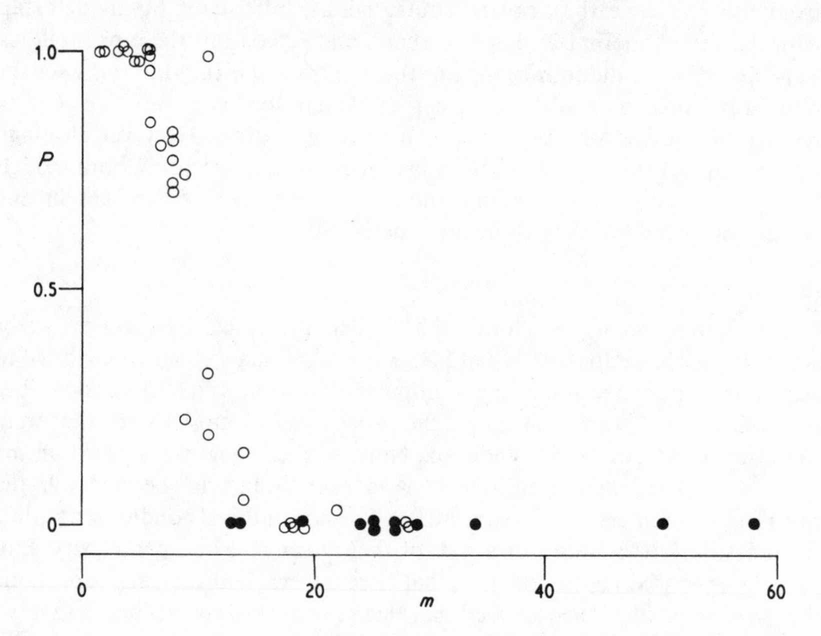

Figure 9.5 The proportion, P, of model food webs that are inverval against the number of predators, m, in the webs. Symbols indicate whether the real web (on which the models are based) is interval (open circles) or not (full circles). From data in Cohen (1978); all versions of all the webs are plotted separately.

(i) The webs that contribute to Cohen's result have large, not small, numbers of predators. For webs with few predators, it is expected and found that the real webs are interval. Webs with many predators are expected to be non-interval and most are: of the 13 different webs (a total of 21 versions) with 12 or more predators, only five (and one other, in one version of web number 11) are interval. For large webs, interval webs are not in the majority, yet they are still more common than would be expected by chance. Indeed, I have calculated that even if only one of these 13 webs were interval it would still be unusual statistically. Two more consequences emerge from this result.

(ii) If extensive lumping of species into 'kinds of organisms' were responsible for Cohen's result, the excess of interval webs would be expected to be caused by the webs with fewer kinds of organisms, namely, those that are the most lumped. The opposite is true, however, for it is the webs with more kinds of predators that give the result. These are the webs that have, by far, the better descriptions of the food webs they represent.

(iii) Finally, is there anything about these five interval, but large, webs that makes them unusual? If so, do their singularities explain Cohen's result? A possibility is suggested by the analysis of Bird's (1930) willow forest web which Cohen used to illustrate his techniques. However, I shall consider the slightly more complex example of Niering's (1963) study of a coral atoll (web 11, Table 9.4); this web can be interval or not, depending on the version. Shown in Fig. 9.6 is what Cohen calls a 'niche overlap graph', in which species that share prey are connected. This web has 19 predators and, using model six, Cohen found that none of his models of this web was interval. The real web is interval if the possible interactions discussed by Niering are omitted, but it is non-interval in the complete version. I find it more interesting that the complete web is also interval if man (species 9) is also omitted. Without man, it is clear why, despite its size, the web must be interval. There are several groups of predators which reasonably could not be expected to share prey extensively.

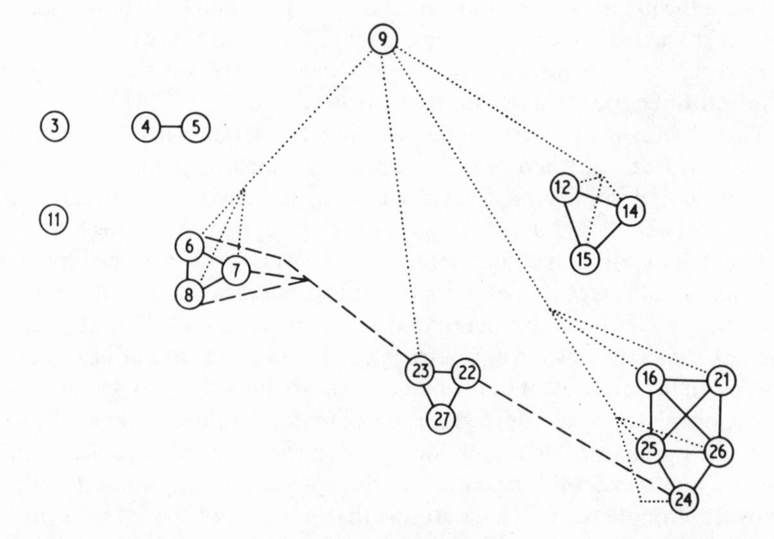

Figure 9.6 A niche overlap graph of a coral atoll food web. Species connected by lines share one or more species of prey. Full lines: species definitely share prey; broken lines: species probably share prey; dotted lines: species share prey with species (9), man. The geometrical relationships between the species are arbitrary. Legend: (3) zooplankton, (4) invertebrates, (5) fish, (6) terns, (7) frigate-birds, (8) boobies, (9) man, (11) sea-turtles, (12) pigs, (14) rats, (15) coconut crabs, (16) fowl, (21) insects, (22) skinks, (23) reef heron, (24) starlings, (25) land crustacea, (26) fungi, snails and annelids, and (27) geckos. Data are from Niering (1963); I have used the same numbers for species as Cohen (1978).

Species 3, 4, 5, 6, 7, 8 and 11 are marine (either invertebrates, 3 and 4, fish, 5, seabirds, 6, 7 and 8, or turtles, 11) and the rest are terrestrial but include species from the atoll's shore (skinks, 22, geckos, 27 and Reef Heron, 23) and from further inland. Individually, these groups – guilds, if you prefer – contain sufficiently few species that it is unlikely, or impossible, for them to be non-interval. The web is interval, despite its size, because it is a composite of several structures, each too simple to be non-interval. The effect of man is to link most of these groups, since man feeds on a wide range of species in different habitats and at many trophic levels. Man's feeding connects these various sets of predators and, not surprisingly, the web becomes non-interval. In short, I suggest that the trophic structuring within a web can make the web interval. Such webs have many predators but their individual structures have few, and such structures are usually interval. This suggestion has been proposed independently by Critchlow & Stearns (in preparation).

(b) Trophic structure

What processes will structure a web in such a way that only small groups of species will share prey? The example I have discussed involves differences between the habitats of sea and land. The groups within the willow forest that Cohen discussed were at different trophic levels. In the willow forest, the herbivores and carnivores had two distinct sets of overlaps; the former involved three species and the latter involved five.

This discussion suggests a test: among the large webs (more than 12 predators) are those where prey overlaps are grouped, interval, but those where they are not, non-interval? Consider first the large webs that are interval. In Table 9.4, webs 1.3, 2.0 and 11 have clear groupings that reflect either the different habitats or the different trophic levels to which their predators belong, or both. Web 5 does not have clear groupings, though the absence of two prey species from the diet of one of the predators would neatly separate the web's overlaps into two groups. Webs 1.1 and 28.1 do not have clear-cut groups. In the latter, 19 of the 28 predators share the same prey species (species 27 in Cohen's table) and this reduces the potential for the non-interval pattern. (For, if all predators share only one prey species, the overlaps are complete and the web is interval.) In contrast to these webs, the large non-interval webs show no groupings or include groups that have many species in them.

To summarize, webs with few predators should usually be interval by chance, as indeed they are. Large webs should be non-interval and, indeed, among large webs the non-interval pattern predominates. Large, interval webs occur usually where biological processes subdivide the patterns of overlaps into groups (guilds, trophic levels, compartments) where there may be too few species per group for the web to be other than interval.

Is Cohen's result a consequence of the patterns of trophic structure I have discussed in previous chapters? Does it simply follow that, if omnivory is rare (Chapter 7), herbivores and carnivores will rarely share prey and, thus, their overlaps will be grouped by trophic levels? Or, because predators in major habitat divisions (compartments; Chapter 8) share few prey, must their overlaps also be grouped? Perhaps, but there are two exceptions to my arguments, namely, webs 1.1 and 28.1. Both webs are interval, involve a large number of species and lack distinct structure. My conclusions are (i) that some of the excess of interval webs in the real world is a consequence of the trophic structure of large webs, but (ii) not all of Cohen's data can be explained this way. Consequently, webs in the real world do appear to be one-dimensional more often than would be expected. As before, more data are needed to resolve the question, but in this case the need for more data is acute: Cohen's result is statistically unusual on the evidence from only two webs.

9.4 SUMMARY

In this chapter I consider three more patterns of interactions within webs:

(i) I examine Cohen's (1977, 1978) assertion that the ratio of the species of prey to predators in webs is 3:4. Under certain circumstances, this excess of predators can cause some theoretical problems, but dynamically stable webs with an excess of predators can be produced quite easily.

Cohen's result is derived from 'community' webs, which are webs where feeding interactions have not been selectively omitted. In 'sink' webs, only the prey of a particular set of species are described and other predators of these prey are ignored. Consequently, this selective omission of predators leads to a much higher prey–predator ratio (of about 3.5:1). The statistically significant excess of predators in community webs may be an artifact. Cohen's results are based on the numbers of 'kinds of organisms'. Many ecologists have a clear antipathy towards plant and invertebrate taxonomy, in that while the kinds of organisms at the top of the food web are usually named to species, those at the base are usually described as 'plants', 'insects', 'phytoplankton', etc. Even when the kinds of organisms at the base of the food web are named to species, a much smaller proportion of the plants in a community, as compared to vertebrates, are shown. The effect of these biases is to reduce selectively the numbers of prey species in a web, and this must lead to an underestimate of the prey–predator ratio. In short, it is far from certain that there are more species of predators than prey in natural communities.

(ii) I examine the correlation between the number of species of prey that a species exploits and the number of predatory species that the species suffers. A positive correlation might be expected between the two variables: a species might require a diversity of species of prey to compensate for the attentions of a variety of predators. In fact, there is a negative correlation between the two variables. Such a negative correlation may be expected *across* trophic levels. Species high in the food web have few predators above them, but their prey are at many trophic levels beneath them. Conversely, species low in the food web have few potential prey but predators at several trophic levels. The expected positive correlation is reasonable for species *along* trophic levels. It is possible that both negative and positive correlations exist within the same web. When all the species in the web are considered, which correlation predominates will depend on whether the web is a 'tall, skinny' one with few species at several trophic levels or a 'short, fat' one with the reverse. The shape of the web recorded in the literature may reflect its true shape but, most likely, it reflects the interests and biases of the observer who recorded it.

Next, I ask: given the shape of a web, what are the expected correlations between the number of species of prey and predators? On average, I find that real webs do not differ statistically from random webs constrained to resemble them in a number of biologically reasonable ways. The observed negative correlations are what would be expected, given the shapes of the webs. So, because the shapes of the webs critically depend on the observers' biases in grouping species at different levels, the overall negative correlation may also be an artifact.

Negative correlations are, however, biologically plausible. Moreover, some well documented and carefully described webs with little species lumping have negative correlations. What are the consequences of a negative correlation if one exists? For species across different trophic levels, this requires that species low in the web, with few prey and many predators, compensate for these disadvantages by exploiting prey with higher equilibrium densities and/or by being more efficient in their exploitation. There is evidence for both: there are (usually) more individuals per species lower in the food chain and ecological efficiencies are also usually greater at lower levels. These arise from familiar patterns of energy flow and transfer. Thus, if a negative correlation exists, it may simply reflect energetic constraints on the web.

A negative correlation reduces the advantages of feeding low in a web. More energy is available low in a food web, and it has been suggested (Chapter 6, hypothesis C) that this is a reason for short food chains. Clearly, the greater numbers of predators on species low in a web will reduce these advantages.

(iii) I consider Cohen's (1978) results that the patterns of species of prey exploited by predators are interval more frequently than would be expected by chance. An interval pattern occurs when the predators' use of prey can be expressed as possibly overlapping segments of a line. The non-interval pattern occurs when the overlaps cannot be so expressed, as exemplified in Fig. 9.4. The interval pattern indicates that the predators' trophic niche is one-dimensional.

Cohen produced sets of model food webs with interactions randomized but subject to different sets of biological constraints. At least one set had assumptions similar to the ones I have used in this and other chapters to develop tests of the likelihood of various food web attributes. Cohen showed that interval webs are far more common than would be expected by chance. He further argued that the results could not be easily dismissed as artifacts.

As the number of predators in a web increases, the chance of a model web being interval remains near unity until about 12 predators per web, when it makes a sharp transition to zero. Thus, below 12 predators per web, most model webs are interval, as are real webs. Above 12 predators, most model webs are non-interval, as, indeed, are real webs. But there are enough large, interval webs to give Cohen his result. The larger webs are usually better documented and have the least lumping of species. This rules out species lumping as a potential explanation of Cohen's result. The observation that a few large, interval webs are, alone, responsible for the result demands the question: is there anything special about these webs? For some, but not all, the answer is 'yes'. Some of the webs have trophic patterns that isolate species overlaps into small groups; each group has too few species to be interval. The resulting web, though large, is still interval. Such trophic groupings can be caused by species belonging to different trophic levels or by compartmenting of species interactions which reflect different habitats. Yet, not all of Cohen's large, interval webs display such groupings and, consequently, part of the statistical excess of interval webs remains unexplained.

10 Food web design: causes and consequences

10.1 INTRODUCTION

Now that I have discussed the patterns of species interactions in food webs, it is time for two questions whose answers require some synthesis: (i) what are the causes of these patterns and (ii) what are their consequences to the functioning of the ecosystems to which the species belong?

To answer the first question, I shall review the patterns discussed in previous chapters and present, briefly, the explanations for them. I shall be concerned, in particular, with two subsidiary questions: (a) is there a single cause of the many food web patterns and (b) is this cause the dynamical constraints on stability?

The answers to question (ii) yield justifications for knowledge of food web structure in addition to those of understanding how structures are formed. I shall show that there are relationships between food web design, how quickly systems respond to perturbations, nutrient cycling and how the biomass at a trophic level varies in response to perturbations. These results are in their earliest stages of development and are not meant to be comprehensive. Rather, my purpose is to make the point that an understanding of food web design is necessary, though not sufficient, for an understanding of ecosystem function.

10.2 CAUSES

10.2.1 A catalogue of patterns

The patterns of food web design discussed in previous chapters are:

(i) A miscellaneous group of patterns that assure biological reality. These include the absence of loops of the kind A eats B eats A, A eats B eats C eats A, etc., and absurdities such as predators without prey and singular systems (e.g. where two top-predators feed exclusively on the same species of prey).

(ii) Webs are not too complex. Specifically, their connectance, C, decreases with the number of species in the web, n, so that the product Cn is less than some constant (Fig. 5.1).

(iii) Food chains are short. Typically, a food web has three or four trophic levels (Fig. 10.1(a)).

(iv) Omnivores are scarce. Typically, food chains have one omnivore per top-predator but not the two or three that are possible in four trophic level systems (Fig. 10.1(b)).

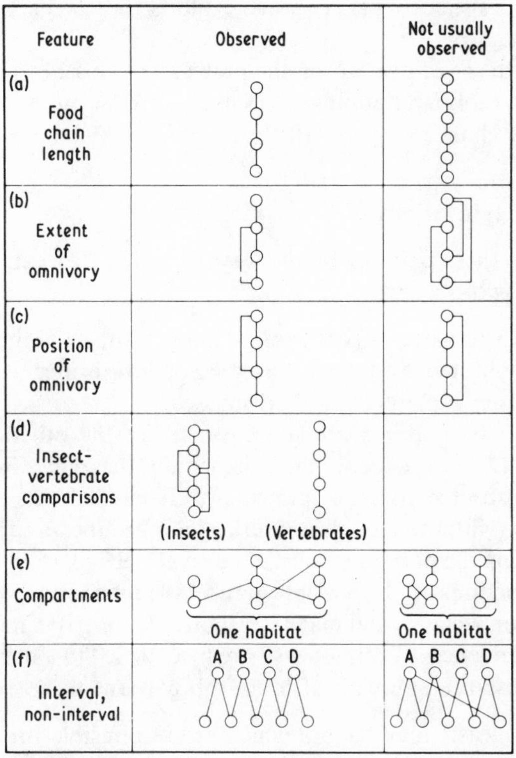

Feature	Observed	Not usually observed
(a) Food chain length		
(b) Extent of omnivory		
(c) Position of omnivory		
(d) Insect-vertebrate comparisons	(Insects) (Vertebrates)	
(e) Compartments	One habitat	One habitat
(f) Interval, non-interval	A B C D	A B C D

Figure 10.1 A partial catalogue of features observed and not observed in real food webs.

(v) Omnivores, when they occur, usually feed on species in adjacent trophic levels (Fig. 10.1 (c)).

(vi) Food webs dominated by insects and their predators and parasitoids have more complex patterns of omnivory than webs dominated by the larger invertebrates and vertebrates. Hence, insect-dominated systems are exceptional to patterns (iv) and (v) (Fig. 10.1(d)).

(vii) There are systems where donor-controlled dynamics are to be expected (e.g. webs with detritivores and scavengers rather than herbivores). These systems also have complex patterns of omnivory (Fig. 10.1(d): substitute detritivores for insects).

(viii) Webs are only occasionally compartmented. Where such compartments exist, they correspond to major habitat divisions (Fig. 10.1(e)).

(ix) Webs are not compartmented within habitats (Fig. 10.1(e)).

(x) The number of species of predators in a web exceeds the number of species of prey (Fig. 9.1).

(xi) The more species of prey a species exploits, the fewer species of predator it suffers (Fig. 9.3(b)).

(xii) The patterns of overlaps in the prey used by a set of predators can be expressed as possibly overlapping intervals along a line more often than would be expected by chance (Fig. 10.1(f)).

10.2.2 Explanations

Before I group these patterns by the ways in which they can be explained, I wish to make two observations:

(a) Two of these patterns (x, xi) are probably artifacts of the ways in which the data were collected. Pattern (vii) is based on little evidence. The remaining patterns, however, are not so easily dismissed.

(b) Some of the patterns are consequences of the others, for there are certainly not 12 independent patterns within the data. For example, if omnivory is restricted (iv) and species in different habitats are in different compartments (viii), then the prey overlaps are grouped and this goes some way in explaining the preponderance of interval webs (pattern xii, Chapter 9). Similarly, restrictions on the possible numbers of interactions between species (i, iv) restrict connectance and may be responsible for pattern (ii). Yet, despite these interdependencies, I cannot easily reduce the list to a mere few patterns; there are a substantial number of food web patterns to be explained.

There are at least three factors which are responsible for these patterns: dynamical constraints, some simple biological considerations, and the physical restrictions intrinsic to the processes of energy flow and transfer.

(a) Dynamical constraints

By far the most important explanation seems to be dynamical constraints. If the species densities in models are required to approach quickly their multispecies equilibrium when perturbed from it, then patterns (i)–(vii) and (ix) are simple consequences. Thus, eight out of the ten probable patterns are explicable in terms of dynamics. Of particular importance is that patterns (ii), (iv)–(vii) and (ix) were *predicted* by theoretical studies of dynamical consequences. A theory that correctly predicts new phenomena is more satisfying than one that merely explains existing patterns. Circularity of argument is much less likely in the former case.

On the other hand, the predominance of dynamics as an explanation may be, in part, a consequence of the emphasis of the current literature on dynamics. That emphasis is, in turn, a reflection of the relative simplicity of the recipe that generates predictions about food web patterns using stability criteria. Were the other explanations to receive so much attention (and were they to produce predictions so clearly), then the list of food web attributes they explain might also be extensive.

Dynamical constraints are not the only ones that explain the patterns just listed. Simple biological realities (pattern (i)) have considerable impact on the model stability. Random webs are much less likely to be stable than models of similar size and complexity without loops, without singularities and without predators without prey (Chapter 4). But to argue that dynamical constraints are the causes of these patterns seems to expand the meaning of stability to the point where it becomes trivial. Biological constraints make it unlikely that two species can be simultaneously each other's predator and prey (a loop). And chance alone must make it improbable that two species with distinct evolutionary histories will exploit exactly the same single species of prey, yet both be devoid of any predators (a singularity). In addition, I have argued that even more subtle patterns (such as (ii) and (vii)) are as (or more!) easily explained by simple biological constraints than by invoking dynamics. I review these arguments in the next section.

None the less, some patterns seem to be best, or necessarily, explained by dynamics (these are patterns (iii)–(vi)). To these may be added the null prediction that food webs should not be compartmented within habitats (ix). Thus, dynamical predictions seem to be solely responsible for about half the patterns observed. Some other patterns (ii, xii) will often follow as a consequence of these. In short, dynamical constraints are sufficient for most but not all of the patterns and necessary for half of them. Dynamical restriction may be the single most important factor in limiting food web designs, but it cannot be the only one.

(b) Biological restrictions

Simple biological knowledge explains some patterns. For instance, a predator must feed on some prey species and, equally apparent, a predator cannot feed on all the species it encounters in its community. A tooth or beak designed for cracking or grinding seeds is not optimally suited for tearing flesh, and vice versa; neither would be optimal for chewing vegetation. With such upper and lower limits on the numbers of species that a predator can exploit, several patterns emerge. Loops are made less likely (pattern (i)); a herbivore, for example, is unlikely to be able to exploit vegetation proficiently and catch a carnivore. That the product of the number of species and connectance is a constant (pattern (ii)) may be another consequence. Thus, as species are added to a community, they will, in part, be of different designs than the ones already present. Consequently, the number of species a predator may exploit will be limited and will not increase linearly with the number of species in the web. If the number of interactions per species is constant, there is a relationship between connectance and the number of species which is indistinguishable from that predicted by dynamics (compare Equations (4.2a, b) and (5.1)). Patterns (viii) and (ix) are also likely to be a consequence of species in different (but not the same) habitats requiring different adaptations. This process, in turn, restricts their feeding to one habitat. Finally, pattern (vii) is most likely to be a consequence of the lack of such restrictions when animals die: if scavengers are rampant omnivores, it is probably because carcasses offer much the same challenges, irrespective of the trophic position of their former inhabitants.

Biological reasons do not seem to be involved in the patterns of omnivory. It is possible that dissimilarity of diet prevents a top-carnivore from feeding on plants in addition to its carnivorous prey (pattern (v)). Yet omnivores that feed on plants and herbivores encounter a comparable problem and are as common as strictly carnivorous omnivores that feed on other carnivores and herbivores. Similarly, although the difficulty of feeding on two trophic levels simultaneously might reduce the number of omnivores down to an average of one per top-predator, it is unclear how this difficulty could explain the relative scarcity of two omnivores per top-predator (pattern (iv)).

In short, four of the patterns may be consequences of biological restrictions on diet, but some other patterns cannot be explained this way.

(c) Energy flow and transfer

Finally, there are energetic constraints. More energy is available at lower trophic levels than at higher ones because as energy is converted from one form to another much of it is lost as heat. This may explain the limitation on food chain lengths, though I express my doubts about this in Chapter 6. It also makes little sense, energetically, for species capable of successfully feeding low in the food chain to exploit species high in the food chain: this is another explanation for the rarity of loops (pattern (i)). Where this does occur – in insectivorous plants, for example – severe nutrient shortages, rather than energy, seem to be the cause.

Unambiguous data may show that species low in a food web have more predators, but fewer prey, than those high in the web (pattern (xi)). If so, the cause of this pattern is likely to be that the greater trophic efficiencies and productivities of species low in the web enable them to withstand the greater number of predators there (Chapter 9).

In short, energetic constraints explain few patterns and none of the probable patterns is explained only by them. Whether this reflects reality or the ecologists' neglect of energetic constraints remains to be seen.

Should all food webs be restricted by the same processes? Obviously not, though some factors are likely to be more important than others. Given the diversity of nature, it is likely that each of these three processes will affect a wide range of food web attributes somewhere and at sometime. But the evidence suggests that some processes – dynamical constraints, for example – are likely to be more widespread in time and space than other processes. The really interesting question is why should a particular food web feature be the result of one process in one ecosystem and another process in another ecosystem? Such a question must await new and better data and more extensive attempts at theoretical synthesis. Currently, the question is premature.

10.3 CONSEQUENCES

10.3.1 Three functions

The aim of the second part of this chapter is, quite simply, to show that the structure of food webs has implications for a number of ecosystem functions. I do not pretend that these implications are fully investigated; they are not. But what is known is sufficient to substantiate my claim of a relationship between food web structure and ecosystem function. The functions I discuss are:

Resilience. How fast do species densities return to equilibrium following a perturbation? I shall use return times (Equation (2.42)) as a measure of resilience.

Nutrient cycling. How tightly do ecosystems retain the nutrients (or pollutants) that make up part of the biomass of their constituent species? Conversely, how quickly are nutrients lost from an ecosystem?

Resistance. How does primary or secondary productivity or the biomass at a particular trophic level change in the face of, say, increased herbivore or carnivore pressure?

The food web structures I shall relate to these functions are food chain length, complexity, and the number of species in the web. Neither the list of functions nor the list of structures is complete. The latter, clearly, should include all the 12 factors discussed in the previous section. To repeat, the aim is to establish some relationships between structure and function, not to document them all.

The simplest strategy would be to explore the nine combinations of the three structures and the three functions. The actual procedure will be more complex in organization but involve fewer relationships. It is represented by the scheme of Fig. 10.2. Several features of this scheme demand attention:

(i) In addition to the variables of structure and function, I have included one of time, namely, succession. Certain kinds of perturbation are sufficiently

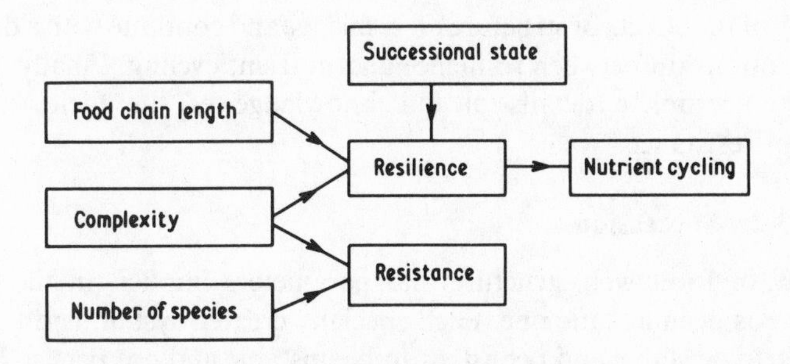

Figure 10.2 A diagram of the causal relationships between food web structures, successional state, and ecosystem functions. This scheme is discussed further in the text.

severe that recovery does not involve merely the species regaining their initial densities. Rather, more or less repeatable sequences of plant and animal species invade the area, persisting for a variety of times. Finally, a *climax* community becomes established that consists of a relatively persistent group of species and usually resembles the system prior to disturbance.

Succession affects both the structural and functional aspects of an ecosystem. Yet, in Fig. 10.2, I have indicated it as affecting only functional attributes. This is deliberate, for I wish to concentrate on the *interaction* between food web structure and succession and their effects on function, rather than on the well known changes in structure (e.g. species number) as succession proceeds.

Why should succession be discussed at all? It is certainly peripheral to the central arguments involving food web structure and ecosystem function. First, the study of successional communities has played the major role in our understanding of the relationship between ecosystem structure and function (Odum, 1969). Succession cannot be excluded if the results on structure and function are to be seen in perspective. I shall present theoretical arguments that succession affects some of the same functions as food web structure and, consequently, both are necessary for an understanding of ecosystem function. Second, the inclusion of succession reveals yet another twist to the complexity–stability problem of earlier chapters.

Succession will be the first topic to be discussed because it affects several of the relationships to follow. But before this there are some other features of Fig. 10.2 to be discussed.

(ii) Not all of the possible relationships between structures and functions are shown. This is usually a reflection of ignorance.

(iii) That there are no direct connections between the structures and nutrient cycling is partly deliberate. The singular position of nutrient cycling in the scheme reflects an important contribution by DeAngelis (1980) which develops links between nutrient cycling and resilience. As I will elaborate presently, resilience does depend directly on food web structure.

In the subsequent discussion, I commence with succession. I proceed with a summary of the effects of structure on resilience and continue with a discussion of the relationship between resilience and nutrient cycling. Finally, I present the very incomplete details of our knowledge of the factors affecting ecosystem resistance.

10.3.2 Succession

The view of food web structure and parameters implicit in the previous chapters has been a static one. Each specially created system had but one of two fates: to be stable and persist, or to be unstable and not persist. Reality is more complex as there are two processes that continually change the structure and parameters of food webs; namely, evolution and succession.

Under pressure from natural selection, species evolve by minimizing the effects of their predators and by becoming more effective predators themselves. This process leads to parameter changes and, at least in geological time, to species extinctions and changes in food web composition. Because of succession, species are continually invading and being lost from communities. Species that invade may or may not be successful and persist. But when they do persist, they alter the shapes of the food webs unless they simply replace existing species. I shall show that, even when species invasions merely effect simple replacement, the functions of the systems are likely to be altered. Similarly, when a species is lost from a system, other losses may follow until a stable system is produced. Such a product is very different in its functions from a comparable, but specially created, system.

I shall not consider the effects of evolution on ecosystem function, but I will present results to show that successional changes affect ecosystem function directly, that is, independently of any indirect effects via structure. Thus, a combined knowledge of successional status and food web structure is required to predict function. Neither structure nor succession is complete in itself.

My reasons for not considering evolutionary changes are:

(i) Little is known about their consequences. What studies are available suggest that the effects are complex and probably not unidirectional. For example, the effect of evolution on the interaction between a predator and its prey may be strongly destabilizing (Schaffer & Rosenzweig, 1978), yet predatory interactions are overwhelmingly common. Are evolutionary old systems less stable than young ones? Are all predator–prey interactions doomed? Are such interactions commonly observed only because they are being created as fast as their self-destruction? Perhaps not. Destabilizing trends may be halted by differential rates of evolution leading to a *coevolutionary steady state* (Rosenzweig, 1973, Rosenzweig & Schaffer, 1978, Schaffer and Rosenzweig, 1978). Alternatively, group selection has been postulated to eliminate predators that are too proficient (Wynne–Edwards 1962, Gilpin, 1975b, Wilson, 1980). The coevolutionary result is mathematically complex; group selection is highly controversial and such results have been obtained only for two-species systems. How evolution affects the functions of multispecies systems and further restricts their possible food web shapes is uncertain. It is likely to remain that way for some time.

(ii) Successional changes may be much faster than evolutionary ones. In succession, the addition of species may take place annually and profoundly affect the structure and function of the system. Moreover, it is across gradients of successional, rather than evolutionary, time that we may wish, or be able, to make comparisons.

'Growing' food webs

There are several studies that seek to understand how food webs produced through some successional process will differ from those specially created.

These studies are all preliminary and have results that may be particularly sensitive to the assumptions of the Lotka–Volterra models used, since the models involve species densities which change considerably over the time course of the simulations. None the less, these studies clearly indicate the importance of successional processes even if they are inadequate to describe the full range of ecologically reasonable possibilities.

The studies of Tregonning and Roberts (Tregonning & Roberts 1978, 1979, Roberts & Tregonning, 1981) have used the mathematical formulations of Roberts' earlier papers (1974; see Chapter 4) combined with a backward elimination of species from successive, unstable equilibria until a feasible and stable equilibrium is obtained. One result was that a sample of webs with n species produced by elimination from $n + m$ species had a higher proportion of stable systems than a sample of specially created n-species webs. The parameters for the $n + m$-species webs and the specially created n-species webs were drawn from the same statistical distributions. Simply, the selection process results in webs that are unusually stable for a given complexity and species number.

An alternative procedure is the stepwise process used by Post & Pimm (in preparation). Starting with six basal species, potential immigrants were successively tested to see if they could invade the existing community and, if invasion was possible, whether the immigrant would displace existing species. Webs were 'grown' in this way for a total of 200 attempted invasions. Different simulations varied both the limits over which the species' parameters were chosen randomly and the probabilities that a species present in the web was the prey or the predator of the immigrant (both were not possible simultaneously). The differences between simulations are relatively minor and peripheral to this book. For convenience, assume that the number of attempted invasions (successful or not) is a measure – certainly not a smooth one – of successional time. Then the cumbersome variable 'number of attempted invasions' can be replaced with the succinct variable 'time', which is also more amenable to field measurement. Two of the time-dependent results of the simulations were:

(i) The number of species in the community increased (accompanied by changes in the food web structure) to some upper limit and remained there (Fig. 10.3(a)).

(ii) The time between successful species invasions continued to increase even after the number of species remained constant (Fig. 10.3(b)).

The first result is much less surprising than the second, for the changes in the numbers of species with time during succession are well known (e.g. Drury & Nisbet, 1973). The second result means that, even for a fixed number of species, there is still a turnover of species: the species number is in dynamic equilibrium; species are continually being lost and gained. This turnover slows with time. Alternatively, each species persists longer the later it enters the

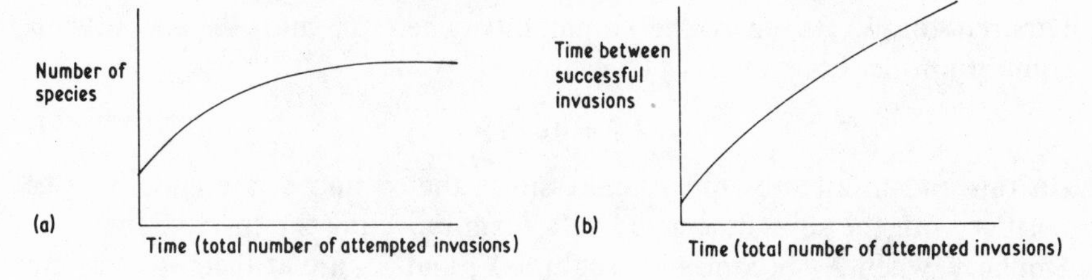

Figure 10.3 A sketch of the results of 'growing' food webs discussed in the text. In the simulations the number of species in the models is asymptotic with time (a), whereas the time between successful invasions continues to increase (b).

community. There are several studies that show this to be a feature of succession in the real world (analyses are provided by Shugart & Hett, 1973).

In short, two functional ecosystem attributes (the resistance of an ecosystem to species invasion and the average persistence times of its constituent species) change with successional time, even after the number of species in the system has reached equilibrium.

Consider an incidental question. There are no studies on the effects of food web structure on species persistence. Do some structures resist invasions and so retain their species longer than others?

The results of Fig. 10.3 are for complex, multispecies models, but their essence can be captured by a simple model. In Fig. 10.4, a species X_3 enters the community as an omnivorous predator of the two existing species X_2, a predator, and X_1, its prey. The invasion results in the extinction of X_2. Now consider another omnivore, X_4: will it be easier or harder for X_4 to repeat X_3's feat of invading the community? With Lotka–Volterra dynamics the condition

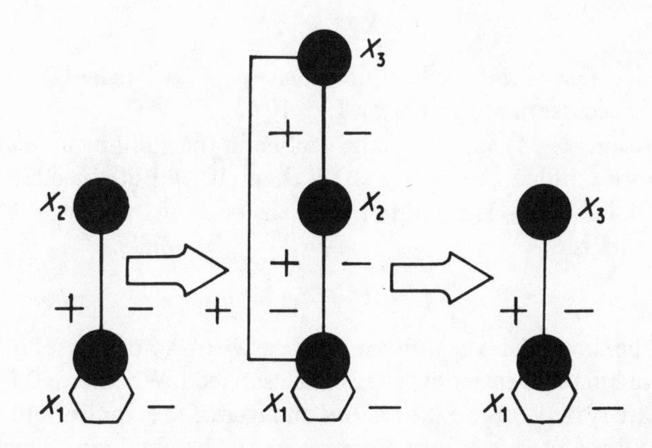

Figure 10.4 A simple successional sequence: X_3 invades and drives X_2 to extinction.

required for X_3 to enter the community when X_1 and X_2 are at their equilibrium densities (X_1^*, X_2^*) is

$$a_{31}X_1^* + a_{32}X_2^* > b_3. \tag{10.1}$$

(In this and in all subsequent expressions, the a_{ij} and b_j are chosen to be positive with the signs associated with them reflecting the form of the web.) Similarly, when X_2 becomes extinct and X_1 and X_3 are at their equilibrium densities (X_1^{**}, X_3^{**}), the condition for X_4 to invade is

$$a_{41}X_1^{**} + a_{43}X_3^{**} > b_4. \tag{10.2}$$

Is it easier or harder to satisfy condition (10.1) than (10.2)? Suppose that the a_{ij} and b_i of the two inequalities are drawn from the same range of possible values. That is, assume that each predator gains, on average, the same advantage from each prey and that each predator declines, on average, at the same rate in the absence of its prey. Under these circumstances, whether condition (10.1) or (10.2) is usually the more restrictive will depend on the relative magnitudes of X_1^* and X_1^{**}, X_2^* and X_3^{**}. Consider the X_1 terms first. For X_2 to be driven to extinction by X_3, $\dot{X}_2 < 0$ when X_1 and X_3 are at their equilibrium densities. This requires that

$$-b_2 + a_{21}X_1^{**} - a_{23}X_3^{**} < 0. \tag{10.3}$$

The term in X_3^{**} will often be relatively small because the equilibrium density of a predator is usually much lower than that of its prey (X_1^{**}). If this is true, then condition (10.3) may be simplified; even if it is not, the following ensures that (10.3) will be true:

$$-b_2 + a_{21}X_1^{**} < 0. \tag{10.4}$$

Now b_2/a_{21} is the equilibrium density of X_1 in the presence of X_2, that is, X_1^*. So, condition (10.4) becomes

$$X_1^{**} < X_1^*. \tag{10.5}$$

Simply, a decrease in the equilibrium density of X_1 will usually accompany the pattern of succession suggested by Fig. 10.4.

In conditions (10.1) and (10.2) the decline in the equilibrium density of X_1 makes it more difficult to satisfy (10.2) than (10.1). But consider the relative sizes of X_2^* and X_3^{**}. The equilibrium density of any predator X_j^* in a two-species model is

$$X_j^* = (b_1 - a_{11}X_1^*)/a_{1j}. \tag{10.6}$$

Thus, for the simple Lotka–Volterra models, when X_1^* decreases X_j^* increases, which would make it easier for (10.2) to be satisfied. Whether (10.1) or (10.2) is easier to satisfy depends on the relative changes of the equilibrium densities of the prey and the predator with the invasion of the omnivore. This is the term a_{11}/a_{1j}, and it is likely to be small. The term is the ratio of what one prey does to its own growth rate divided by the effect that one predator has on the prey's growth rate. When this ratio is small, the increase of X_2^* to X_3^{**} will be small relative to the decrease between X_1^* and X_1^{**}.

In short, it should be harder for a predator to invade the community the more often such invasions have occurred. The parameters required for a species to invade a community and replace its existing predator change each time this event takes place. And the parameters change in such a way that those required for successful invasion become more 'special': other things being equal, the constraints on them will be harder to satisfy. This result confirms the simulation results for more complex models: as succession proceeds, even though the structure of the food web may not change, its properties will. The system becomes harder to invade and consequently the species persist through more attempted invasions.

Another change found by Post & Pimm's simulations involved resilience. Systems were found to decrease in their resilience (return times increased) with succession. Again, this trend continued after the number of species stabilized. The reasons for this trend are complex and, for simple models, it is possible for resilience to increase before decreasing during succession. The trends in the population densities of the autotrophs were towards smaller equilibrium densities in all the simulations, as in the example worked above. Recall that the eigenvalues of a matrix sum to the sum of the diagonal elements of the matrix. For Lotka–Volterra models, the diagonal elements are $a_{ii} X_i^*$, and so as the X_i^* become smaller, the sum of the eigenvalues must approach zero, which means that the resilience of the system decreases.

10.3.3 Resilience

The effects of food web structure on resilience are illustrated in Fig. 10.5. The results are from a wide range of six- and eight-species models (see Fig. 4.6) grouped, respectively, into three and four trophic levels. The models also differ in their connectances: between zero and six additional predator–prey interactions were added to those of the basic models of simple food chains. As the connectance of these models increased, the percentage of the models that were stable decreased rapidly (Fig. 4.7, Table 8.1). In Fig. 10.5, however, I have considered only those models that were stable and, for these, recorded the percentages of each model structure that had return times less than the arbitrary value of 200. This gives a measure of resilience for each model. In the figure, two features are apparent (and statistically significant; Pimm, 1979b):

(i) As discussed in Chapter 6, the models with three trophic levels are more resilient than those with four.

(ii) The more complex a food web is, the more resilient it is, provided that the model is stable in the first place.

Two conclusions emerge from these results. The first is the obvious one that

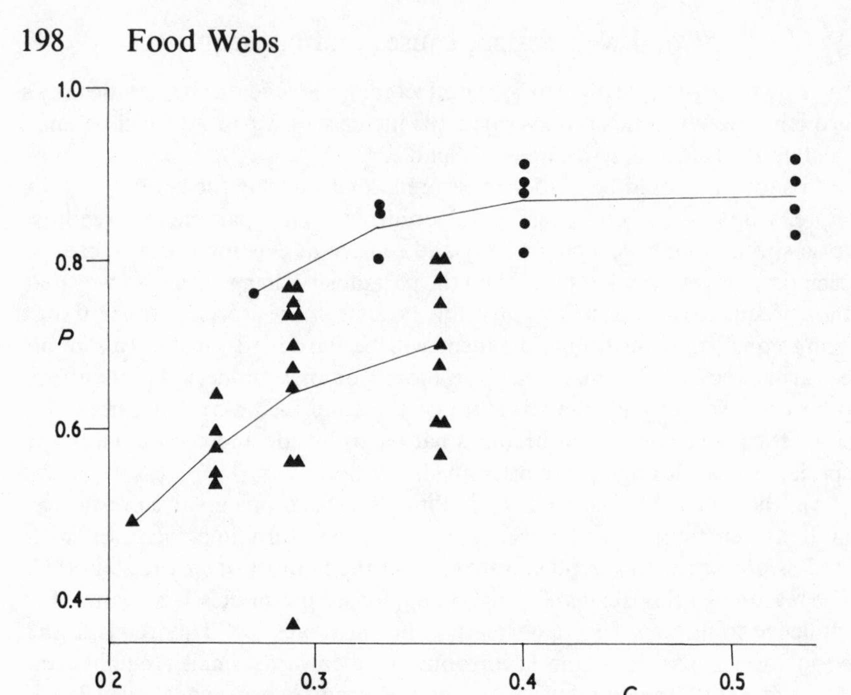

Figure 10.5 The proportion, P, of stable food web models with short return times (< 200) – a measure of system resilience – against the connectance, C, of three trophic level models (circles) and four trophic level models (triangles). The units of return times are discussed on page 28.

an ecosystem function – resilience – is clearly dependent, at least in part, on two features of food web structure. The second involves, yet again, the topic of complexity and stability.

Complexity and functional aspects of stability

In Chapter 4, I asked which of two food webs was more likely to be stable. I pointed out the difficulties with this question. In the sense that the constituent species of both webs persist, both webs could be considered stable. This motivated two related, but answerable questions:

(i) Does stability impose upper or lower limits on complexity?
(ii) Is one system relatively more stable than another?

The first question was discussed in Chapters 4, 5, and 7. I concluded that complex systems are not as likely to be stable as simple ones. Now, for the second question, an apparent paradox is encountered.

Resilience is one reasonable definition of relative stability. The results of Fig. 10.5 show that among those stable systems observed, the more complex systems are relatively more stable. Thus, whether complexity and stability are positively or negatively correlated depends entirely on which one of the two questions is asked.

The theoretical result, that among stable systems those that are complex are relatively more stable than those that are simple, is interesting in other contexts. First, it confirms the early ideas on complexity and stability. More importantly, the result seems eminently suited to field testing: I suggest ways of measuring resilience in Chapter 3; estimating connectance from a food web is straightforward.

Next, recall the result from the previous section that, during succession, species persistence increases. Thus, if persistence is considered the measure of stability, then stability will be greater in later successional stages. This is exactly what Odum (1969) suggested. Incidentally, later successional stages also have more species than earlier ones and diversity is another common measure of complexity.

There is also at least a theoretical relationship between resilience and the patterns of nutrient cycling, as I show next. There should be relationships between food web structure, successional state, and nutrient cycling, which is another assertion of Odum's (1969) paper.

To summarize, stability probably restricts those systems observed to relatively simple structures. But, within the systems observed, theoretical studies suggest that complex systems are more resilient than simple ones. Finally I suggest that, across a successional gradient, later stages have species that persist longer, but will be less resilient, than those in earlier stages.

10.3.4 Nutrient cycling

In an important paper, DeAngelis (1980) established a close theoretical relationship between resilience and nutrient cycling. Now, atoms of a given element may be held tightly by a system and may be recycled many times before they leave the system. DeAngelis' theory is simple: the shorter the time a unit of material resides in an ecosystem, the more resilient the ecosystem will be to perturbations of that material. The theory was motivated in part by a study (Jordan, Kline & Sasscer, 1972) which found that perturbations to non-essential elements in three forest ecosystems disappeared faster than ones to essential elements. Jordan *et al.* argued that essential elements, such as calcium and phosphorous, tend to be tightly cycled. Perturbations to these elements should dampen only slowly, whereas such non-essential elements as cesium are lost quickly and perturbations should disappear rapidly.

DeAngelis developed his theoretical arguments in two ways: by analyses of simple three-compartment models, and by computer simulations of models too complex for analysis. His three-compartment model involved transfers into compartment one, from one to two, from two to three and recycling from three to one. There were losses from all three compartments. One way of viewing this model is for the compartments to be: one, soil; two, plants; and three, herbivores and detritivores through which nutrients must pass before returning to the soil.

When recycling is strong, fluxes out of a system are small compared with fluxes from one compartment to another. DeAngelis called the ratio of these quantities ε, and showed that the largest eigenvalue of the system varies approximately as $-\varepsilon$. He then defined an index of recycling, R, that measured the average number of times a unit of material was recycled before leaving the system. With tight recycling R is approximately $1/3\varepsilon$ for the three-compartment model. Thus, long return times ($1/\varepsilon$) vary in the same way as this recycling index ($1/3\varepsilon$): perturbations to tightly recycled nutrients should disappear only slowly, and vice versa. For a large series of simulations of more complex models (including those four-species, four trophic level models of Fig. 7.2), DeAngelis demonstrated a close positive correlation between return times and another index of recycling, namely, transit times (DeAngelis, 1980, personal communication). Transit time is the mean time it takes for a unit of nutrient to leave the system from when it first entered. The index depends on several factors, but the more tightly a nutrient is recycled, the longer it will take to leave the system.

Food web design and nutrient cycling

There are three results to be discussed. First, food web structure directly affects ecosystem resilience. Second, resilience is also modified by successional processes, even when food web structure has stabilized. Finally, resilience and the rapidity of nutrient losses are correlated. From these results, it is likely that food web structure has a simple effect on nutrient cycling, though this is not certain. DeAngelis' results are for systems with fixed structures, and he has shown that resilience and recycling are related within, and not between, those structures. His results and those on the changes of resilience during succession are, therefore, both within fixed structures. Consequently, theoretical predictions about the change of patterns of recycling during succession emerge with some certainty. If, as I suggested earlier, resilience decreased with succession, then nutrients should be more tightly held and recycled as succession proceeds. This is exactly what Odum (1969) suggested.

Current work (DeAngelis, personal communication) is concerned with the investigation of whether resilience and recycling are directly related: that is, what are their relationships *across* a range of food web structures? Two possibilities from these studies are: (i) resilience *per se* may be shown to be related to recycling and food web structure may correlate with recycling only in that it indirectly affects resilience; and (ii) different food web structure may modify the exact relationship between resilience and nutrient cycling. The overall conclusion for both of these would be the same, namely, the features of nutrient cycling depend, in part, on food web design.

10.3.5 Resistance

Recall the idea introduced in Chapter 5 when I discussed McNaughton's (1977) experimental results. McNaughton experimentally investigated the effects of an African Buffalo's grazing of both species-poor and species-rich grasslands. The grazing changed species diversity in the latter plot more than in the former, and this is consistent with the idea that increasing numbers of species diminish the stability of a system. McNaughton argued the reverse by presenting evidence to show that the plant biomass was reduced, proportionately, six times more in the species-poor than in the species-rich plots. He argued that the more plant species there were in an area, the more likely at least one would be able to avoid grazing and so compensate for the loss in the biomass of the other species. In the species-rich plots, one species (*Thimeda triandra*) was identified which increased in abundance dramatically when the Buffalo was present. *Thimeda*, by resisting the effects of grazing, simultaneously compensated for the effects of grazing on the other species and, by becoming numerically dominant, contributed to the reduction in species diversity. McNaughton's results are convincing evidence that the answer to functional aspects of the stability–complexity question may be entirely different from the structural ones posed in Chapters 4 and 5.

The processes described by McNaughton have not attracted theoretical treatment. Consider species competing for resources along a gradient with a capricious environment selecting only a small part of that gradient at any instant. How should the variability in total species biomass change with increasing numbers of species and changing degrees of species polyphagy? How should species number and prey overlap affect the variation in the total biomass of a set of predators subjected to variations in which species of prey were present? The reverse problem of the consequences to the total biomass of prey, given variations in which of their predators were present, is directly comparable to McNaughton's experiments and is also worthy of study. Some preliminary work on these questions has been undertaken by Juha Tuomi and co-workers (Tuomi & Niemela, 1979, Tuomi, in preparation), but here is clearly a potentially rich area for theoretical and experimental studies.

10.4 SUMMARY

I have discussed 12 patterns of food web design and their possible explanations. Two of the patterns are likely to be artifacts of the ways in which the data were collected and for another there is little certain evidence. Dynamical constraints, simple biological restrictions, and the limitations on energy flow explain varying numbers of these patterns. The results are summarized in Table 10.1.

Finally, I consider the consequences of three of these structural patterns to food web function. The three structural characteristics are species number, the

Table 10.1 A summary of the causes of food web patterns.

Pattern number	*Pattern*	*Is the pattern uncertain?*	*Is the pattern explained by:* Dynamics	Biology	Energy
(i)	no loops, predators without prey, singular systems		yes	yes	yes
(ii)	limits on complexity		yes	yes	
(iii)	food chain lengths are limited		yes		perhaps
(iv)	omnivores are scarce		yes		
(v)	omnivores feed on adjacent trophic levels		yes		
(vi)	insects and their parasitoids are exceptional to patterns 4 and 5		yes		
(vii)	scavengers feed on many tropic levels	yes	yes	yes	
(viii)	Compartments correspond to habitat divisions			yes	
(ix)	webs are not compartmented within habitats		yes	yes	
(x)	more predators than prey	yes			
(xi)	the more species of prey a species exploits, the fewer species of predators it suffers	yes			yes
(xii)	prey overlaps are interval		in part (from (iv))	in part (from (viii))	

number of trophic levels, and complexity; the three functions are resilience, nutrient cycling, and resistance. Early studies on the relationships between structure and function have stressed changes during succession. Consequently, I present briefly some theoretical results on succession. They show that, during succession, there are increases in the resistance of a system to species invasion even after the number of species in the system has attained equilibrium. Resilience, however, seems to decrease through succession and also depends on at least two structural features, namely, complexity and the number of trophic levels. Thus the resilience of a system depends critically on its structure, but even among identical structures resilience will vary with successional state.

Nutrient cycling is related to resilience. When nutrients are tightly recycled, perturbations to the nutrient levels disappear only slowly, and vice versa. Whether the tightness of nutrient cycling directly depends on structure is not yet known. But there is likely to be at least an indirect effect: structure affects resilience and resilience affects nutrient cycling.

How the total biomass of the species along on trophic level varies as the densities of its resources, prey or predators change, is the subject of some experimental and theoretical studies. McNaughton (1977) found that a herbivore had less effect on the total biomass of plants in a species-rich than in a species-poor grassland community.

Although the lists of structures and functions are far from complete two conclusions emerge:

(i) The functioning of an ecosystem depends, partly, on the structure of its food web. The relative importance of structure, successional state, and other unspecified factors remains to be investigated.

(ii) Several of these results relate to the problems of complexity and stability. Although stability restricts the systems observed to those that are relatively simple, among those systems observed the more complex will be the more resilient. However, late successional stages with more species (another measure of complexity) may be less resilient than early stages with few species. And there is good observational and theoretical evidence that late successional stages resist species invasions and retain their species longer than early stages. In short, the answer to the complexity–stability question depends entirely on which one of the many possible questions is being asked.

Bibliography

Askew, R. R. (1961), On the biology of inhabitants of Oak galls of Cynipidae (Hymenoptera) in Britain. *Transactions of the Society for British Entomology*, **14**, 237–69.

Askew, R. R. (1971), *Parasitic Insects*, Heinemann, London.

Batten, L. A. & Marchant, J. H. (1976), Bird population changes for the years 1973–1974. *Bird Study*, **23**, 11–20.

Battan, L. A. & Marchant, J. H. (1977), Bird population changes for the years 1974–1975. *Bird Study*, **24**, 159–64.

Beaver, R. A. (1979), Fauna and foodwebs of pitcher plants in West Malaysia. *The Malayan Nature Journal*, **33**, 1–10.

Beddington, J. R., Free, C. A. & Lawton, J. H. (1975), Dynamic complexity in predator-prey models framed in difference equations. *Nature*, **225**, 58–60.

Beddington, J. R., Free, C. A. & Lawton, J. H. (1978), Modelling biological control: on the characteristics of successful natural enemies. *Nature*, **273**, 513–9.

Beddington, J. R. & Hammond, P. S. (1977), On the dynamics of host–parasitoid–hyperparasitoid interactions. *Journal of Animal Ecology*, **46**, 811–22.

Bellows, T. S. (1981), The descriptive properties of some models for density dependence. *Journal of Animal Ecology*, **50**, 139–56.

Bird, R. P. (1930), Biotic communities of the aspen parkland of central Canada. *Ecology*, **11**, 356–442.

Birkeland, C. (1974), Interactions between a sea pen and seven of its predators. *Ecological Monographs*, **44**, 211–32.

Blindloss, M. E., Holden, A. V., Bailey-Watts, A. E. & Smith, I. R. (1972), Phytoplankton production, chemical and physical conditions in Loch Leven. *Productivity Problems of Freshwaters* (ed. Z. Kazak & A. Hillbricht-Ilkowska), Polish Scientific Publishers, Warsaw.

Bliss, L. C. (ed.) (1977), *Truelove Lowland, Devon Island, Canada: A High Arctic Ecosystem*, University of Alberta Press, Edmonton.

Bunt, J. S. (1975), Primary productivity of marine ecosystems. *Primary Productivity of the Biosphere* (ed. H. Leith & R. H. Whittaker), Springer-Verlag, New York.

Burgis, M. J., Darlington, J. P. E. C., Dunn, I. G., Ganf, G. G., Gwahaba, J. J. & McGowan, L. M. (1973), The biomass and distribution of organisms in Lake George, Uganda. *Proceedings of the Royal Society*, B, **184**, 271–98.

Clarke, T. A., Flechsig, A. O. & Grigg, R. W. (1967), Ecological studies during Sealab II. *Science*, **157**, 1381–9.

Cohen, J. E. (1977), Ratio of prey to predators in community webs. *Nature*, **270**, 165–7.

Cohen, J. E. (1978), *Food Webs and Niche Space. Monographs in Population Biology*, **11**, Princeton University Press, Princeton, NJ.

Connell, J. H. (1975), Some mechanisms producing structure in natural communities.

Ecology and Evolution of Communities (ed. M. L. Cody & J. M. Diamond), Harvard University Press, Cambridge, Mass.

Damuth, J. (1981), Population density and body size in mammals. *Nature*, **290**, 699–702.

Dart, J. K. G. (1972), Echinoids, algal lawn and coral recolonisation. *Nature*, **239**, 50–1.

Dayton, P. K. (1975a), Experimental studies of algal canopy interactions in a sea-otter dominated kelp community at Amchitka Island, Alaska. *Fisheries Bulletin*, **73**, 230–7.

Dayton, P. K. (1975b), Experimental evaluation of ecological dominance in a rocky intertidal algal community. *Ecological Monographs*, **45**, 147–59.

DeAngelis, D. L. (1975), Stability and connectance in food web models. *Ecology*, **56**, 238–43.

DeAngelis, D. L. (1980), Energy flow, nutrient cycling and ecosystem resilience. *Ecology*, **61**, 764–71.

Draper, N. R. & Smith, H. (1966), *Applied Regression Analysis*, Wiley, New York.

Drury, W. H. & Nisbet, I. C. T. (1973), Succession. *The Arnold Arboretum Journal*, **54**, 331–68.

Dykyjova, D. & Kvet, J. (1978), *Pond Littoral Ecosystems*, Springer-Verlag, New York.

Elton, C. S. (1927), *Animal Ecology*, Macmillan, New York.

Elton, C. S. (1958), *The Ecology of Invasions by Animals and Plants*, Chapman & Hall, London.

Elton, C. S. (1966), *The Pattern of Animal Communities*, Chapman & Hall, London.

Elton, C. S. & Nicholson, M. (1942), The ten-year cycle in numbers of the Lynx in Canada. *Journal of Animal Ecology*, **11**, 215–44.

Estes, J. A., Smith, N. S. & Palmisano, J. L. (1978), Sea-otter predation and community organization in the western Aleutian Islands, Alaska. *Ecology*, **59**, 882–933.

Fisher, R. A. (1950), *Statistical Methods for Research Workers*, 11th Edition, Oliver & Boyd, Edinburgh.

Flanders, H., Korfhage, R. R. & Price, J. P. (1970), *Calculus*, Academic Press, London.

Force, D. C. (1974), Ecology of insect host-parasitoid communities. *Science*, **184**, 624–32.

French, N. R. (ed.) (1979), *Perspectives in Grassland Ecology*, Springer-Verlag, New York.

Fryer, G. (1957), The trophic interrelationships and ecology of some littoral communities of Lake Nyasa with special reference to the fishes and a discussion of the evolution of a group of rock-dwelling Cichlidae. *Proceedings of the Zoological Society of London*, **132**, 153–281.

Futuyma, D. J. & Gould, F. (1979), Associations of plants and insects in a deciduous forest. *Ecology*, **49**, 33–50.

Gard, T. C. & Hallam, T. G. (1979), Persistence in food webs, I: Lotka–Volterra food chains, *Bulletin of Mathematical Biology*, **41**, 877–91.

Gardarsson, A. (1979), Waterfowl populations of Lake Myvatn and recent changes in numbers and food habits. *Oikos*, **32**, 250–70.

Gardner, M. R. & Ashby, W. R. (1970), Connectance of large, dynamical (cybernetic) systems: critical values for stability. *Nature*, **228**, 784.

Gibson, C. W. D. (1976), The importance of food plants for the distribution and abundance of some Stenodemini (Heteroptera; Miridae) of limestone grassland. *Oecologia*, **25**, 55–76.

Gilpin, M. E. (1975a), Stability of feasible predator-prey systems. *Nature*, **254**, 137–39.

Gilpin, M. E. (1975b), *Group Selection in Predator-Prey Communities. Monographs in Population Biology*, **9**, Princeton University Press, Princeton, New Jersey.

Glynn, P. W. (1976), Some physical and biological determinants of coral community structure in the Eastern Pacific. *Ecological Monographs*, **46**, 431–56.

Goh, B. S. (1975), Stability, vulnerability and persistence of complex ecosystems. *Ecological Modelling*, **1**, 105–16.

Goh, B. S. (1976), Nonvulnerability of ecosystems in unpredictable environments. *Theoretical Population Biology*, **10**, 83–95.

Goh, B. S. (1977), Global stability in many species systems. *The American Naturalist*, **111**, 135–43.

Goh, B. S. & Jennings, L. S. (1977), Feasibility and stability in randomly assembled Lotka–Volterra models. *Ecological Modelling*, **3**, 63–71.

Goldman, C. R., Mason, D. T. & Hobbie, J. E. (1967), Two Antarctic desert lakes. *Limnology and Oceanography*, **12**, 295–310.

Goodman, D. (1975), The theory of diversity–stability relationships in ecology. *Quarterly Review of Biology*, **50**, 237–66.

Grove, W. E. (1966), *Brief Numerical Methods*, Prentice Hall, Englewood Cliffs, NJ.

Haedrich, R. L. & Rowe, G. T. (1977), Megafaunal biomass in the deep sea. *Nature*, **269**, 141–2.

Hall, D. J. Cooper, W. E. & Werner, E. E. (1970), An experimental approach to population dynamics and structure of freshwater animal communities. *Limnology and Oceanography*, **15**, 839–928.

Hallam, T. G. (1980), Effects of cooperation on competitive systems. *Journal of Theoretical Biology*, **82**, 415–23.

Hallam, T. G., Svoboda, L. J. & Gard, T. C. (1979), Persistence and extinction in three species Lotka–Volterra competitive systems. *Mathematical Biosciences*, **46**, 117–24.

Hansen, R. M. & Uekart, D. N. (1970), Dietary similiarity of some primary consumers. *Ecology*, **51**, 640–8.

Hardy, A. C. (1924), The herring in relation to its animate environment. Part 1: The food and feeding habits of the herring with special reference to the East Coast of England. *Fisheries Investigation*, Series II, **7**, 1–45.

Harper, J. L. (1969), The role of predation in vegetation diversity. *Cold Spring Harbor Symposia*, **22**, 48–62.

Harris, J. R. W. (1979), The evidence for species guilds is an artefact. *Nature*, **279**, 350–1.

Harrison, G. W. (1980), Global stability of food chains. *The American Naturalist*, **114**, 455–7.

Harrison, J. L. (1962), The distribution of feeding habits among animals in a tropical rain forest. *Journal of Animal Ecology*, **31**, 53–63.

Hassell, M. P. (1979), *The Dynamics of Arthropod Predator–Prey Systems. Monographs in Population Biology*, **13**, Princeton University Press, Princeton, NJ.

Hastings, H. H. & Conrad, M. (1979), Length and evolutionary stability of food chains. *Nature*, **282**, 838–9.

Holt, R. D. (1977), Predation, apparent competition and the structure of prey communities. *Theoretical Population Biology*, **11**, 197–229.

Humphries, W. F. (1979), Production and respiration in animal populations. *Journal of Animal Ecology*, **48**, 427–54.

Hurd, L. E., Mellinger, M. V., Wolf, L. L. & McNaughton, S. J. (1971), Stability and diversity at three trophic levels in terrestrial successional ecosystems. *Science*, **173**, 1134–6.

Hurlbert, S. H., Mulla, M. S. & Willson, H. R. (1972), Effects of an organophosphate insecticide on the phytoplankton, zooplankton and insect populations of freshwater ponds. *Ecological Monographs*, **42**, 269–99.

Hutchinson, G. E. (1959), Homage to Santa Rosalia or why are there so many kinds of animals? *The American Naturalist*, **93**, 145–59.

Hutchinson, G. E. & MacArthur, R. H. (1959), A theoretical ecological model of size distributions among species of animals. *The American Naturalist*, **93**, 117–26.

Ito, Y. (1972), On the methods for determining density dependence by means of regression. *Oceologia*, **10**, 347–72.

Jansson, A. M. (1967), The food web of the *Cladophora*–belt fauna. *Helgoländer wissenshaftliche Meeresuntersunchungen*, **15**, 571–588.

Jeffries, C. (1974), Qualitative stability and digraphs in model ecosystems. *Ecology*, **55**, 1415–9.

Joern, A. (1979), Feeding patterns in grasshoppers (Orthoptera; Acrididae); factors affecting specialization. *Oecologia*, **38**, 325–48.

Johanssen, P. M. (ed.) (1979), Ecology of eutrophic, subarctic Lake Myvatn and the River Laxa. *Oikos*, **32**, 1–308.

Jones, J. R. E. (1959), A further ecological study on calcareous streams in the 'Black Mountain' region of South Wales. *Journal of Animal Ecology*, **28**, 142–59.

Jones, N. S. (1948), Observations and experiments on the biology of *Patella vulgata* at Port St Mary, Isle of Man. *Proceedings and Transactions of the Liverpool Biological Society*, **56**, 60–7.

Jordan, C. F., Kline, J. R. & Sasscer, D. S. (1972), Effective stability of mineral cycles in forest ecosystems. *The American Naturalist*, **106**, 237–53.

Kitching, J. H. & Ebling, F. J. (1961), The ecology of Loch Ine. *Journal of Animal Ecology*, **30**, 373–83.

Kitching, R. L. (1981), Community structure in water-filled tree-holes in Europe and Australia – some comparisons and speculations. *Phytotelmata: Terrestrial Plants as Hosts of Aquatic Insect Communities* (ed. H. Frank & P. Lounibos), Plexus Press, Marlton, New Jersey.

Knox, E. A. (1970), Antarctic marine ecosystems. *Antarctic Ecology*, (ed. M. W. Holdgate), Academic Press, London.

Koblentz-Mishke, O. J., Volkovinsky, V. V. & Kabanova, J. G. (1970), Plankton primary productivity of the world ocean. *Scientific Exploration of the South Pacific* (ed. W. S. Wooster), National Academy of Sciences, Washington, DC.

Koepcke, H. W. & Koepcke, M. (1952), Sobre el proceso de transformacion de la materia organica en las playas arenosas marinas del Peru. *Publicaciones de Universidad Nacional Mayor de San Marcos, Zoologie, Serie A*, **8**.

Kowlaski, R. (1977), Further elabouration of the winter moth population model. *Journal of Animal Ecology*, **46**, 471–82.

Kruuk, H. (1972), *The Spotted Hyena*, The University of Chicago Press, Chicago.

Lack, D. L. (1966), *Population Studies of Birds*, Oxford University Press, Oxford.

Lack, D. L. (1971), *Ecological Isolation in Birds*, Blackwell Scientific Publications, Oxford.

Lakatos, I. (1978), *The Methodology of Scientific Research Programmes* (*Philosophical Papers of Imre Lakatos*) (ed. J. Worrall & G. Guthrie), Cambridge University Press, Cambridge.

Lamotte, M. (1975), The structure and function of a tropical savannah ecosystem. *Tropical Ecological Systems: Trends in Terrestrial and Aquatic Research* (ed. F. R. Golley & E. Medina), Springer-Verlag, New York.

Langston, W. (1981), Pterosaurs. *Scientific American*, **244**, 122–37.

Larsson, P., Brittain, J. E., Lein, L., Lillehammer, A. & Tangen, K. (1978), The lake ecosystem of Øvre Heimdalsvatn. *Holarctic Ecology*, **1**, 304–20.

LaSalle, J. P. (1960), The extent of asymptotic stability. *Proceedings of the National Academy of Sciences U.S.A.*, **46**, 363–365.

Lawlor, L. R. (1978), A comment on randomly constructed ecosystem models. *The American Naturalist*, **112**, 445–7.

Lawton, J. H. (1974), The structure of the arthropod community on bracken (*Pteridium aquilinum* (L.) Kuhn). *Biology of Bracken*, (ed. F. H. Perring), Academic Press, New York.

Lawton, J. H. & McNeill, S. (1979), Between the devil and the deep blue sea: on the problems of being an herbivore. *Population Dynamics. Symposia of the British Ecological Society*, **20** (ed. R. M. Anderson, B. D. Turner & L. R. Taylor), Blackwell Scientific Publications, Oxford.

Lawton, J. H. & Pimm, S. L. (1978), Population dynamics and the length of food chains. *Nature*, **272**, 189–90.

Lawton, J. H. & Price, P. W. (1979), Species richness of parasites on hosts: Agromyzid flies on the British Umbelliferae. *Journal of Animal Ecology*, **48**, 619–38.

Lawton, J. H. & Rallison, S. P. (1979), Stability and diversity in grassland communities. *Nature*, **279**, 351.

Lawton, J. H. & Schroder, D. (1977), Effects of plant type, size and of geographical range and taxanomic isolation on the number of insect species associated with British plants. *Nature*, **265**, 137–40.

Leighton, W. (1970), *Ordinary Differential Equations*, 3rd Edn, Wadsworth, Belmont.

Leith, H. (1975), Primary productivity of the major vegetation units of the world. *Primary Productivity of the Biosphere* (ed. H. Leith & R. H. Whittaker), Springer-Verlag, New York.

Likens, G. E. (1975), Primary productivity of inland aquatic ecosystems. *Primary Productivity of the Biosphere* (ed. H. Leith & R. H. Whittaker), Springer-Verlag, New York.

Lindemann, R. L. (1942), The trophic-dynamic aspect of ecology. *Ecology*, **23**, 399–413.

Lotka, A. J. (1925), *Elements of Physical Biology*, Willliams and Wilkins, Baltimore, Md.

Luckinbill, L. S. & Fenton, M. (1978), Regulation and environmental variability in experimental populations of Protozoa. *Ecology*, **59**, 1271–6.

Lynch, M. (1979), Predation, competition and zooplankton community structure: an experimental study. *Limnology and Oceanography*, **24**, 253–72.

MacArthur, R. H. (1955), Fluctuations of animal populations and a measure of community stability. *Ecology*, **36**, 533–6.

McMurtrie, A. (1975), Determinants of stability of large, randomly connected systems. *Journal of Theoretical Biology*, **50**, 1–11.

McNaughton, S. J. (1977), Diversity and stability of ecological communities; a comment on the role of empiricisms in ecology. *The American Naturalist*, **111**, 515–25.

McNaughton, S. J. (1978), Stability and diversity of ecological communities. *Nature*, **274**, 251–3.

McNeill, S. & Lawton, J. H. (1970), Annual production and respiration in animal populations. *Nature*, **225**, 472–4.

Mann, K. H. (1964), The case history: the river Thames. *River Ecology and Man* (ed. R. T. Oglesby, C. A. Carlson & J. A. McCann), Academic Press, New York.

Mann, K. H. (1965), Energy transformations by a population of fish in the river Thames. *Journal of Animal Ecology*, **34**, 253–75.

Mann, K. H., Britton, R. H., Kowalczewski, A., Lack, T. J., Mathews, C. P. & McDonald, I. (1972), Productivity and energy flow at all trophic levels in the river Thames. *Productivity Problems of Freshwaters* (ed. Z. Kazak & A. Hillbricht-Ilkowska), Polish Scientific Publishers, Warsaw.

Marchant, J. H. (1978), Bird population changes for the years 1975–1976. *Bird Study*, **25**, 245–52.

Margalef, R. (1975), Diversity, stability and maturity in natural ecosystems. *Unifying Concepts in Ecology* (ed. W. H. van Dobben & R. H. Lowe-McConnell), Junk, The Hague.

Marples, T. G. (1966), A radionuclide tracer study of arthropod food chains in a *Spartina* salt-marsh ecosystem. *Ecology*, **47**, 270–7.

May, R. M. (1972), Will a large complex system be stable? *Nature*, **238**, 413–4.

May, R. M. (1973a), Qualitative stability in model ecosystems. *Ecology*, **54**, 638–41.

May, R. M. (1973b), *Stability and Complexity in Model Ecosystems. Monographs in Population Biology*, **6**, Princeton University Press, Princeton, NJ.

May, R. M. (1974a), Biological populations with non-overlapping generations; stable points, stable cycles and chaos. *Science*, **186**, 647–8.

May, R. M. (1974b), On the theory of niche overlap. *Theoretical Population Biology.*, **5**, 297–332.

May, R. M. (1975), Stability in ecosystems: some comments. *Unifying Concepts in Ecology*. (ed. W. H. van Dobben & R. H. Lowe-McConnell), Junk, The Hague.

May, R. M. (1979), The structure and dynamics of ecological communities. *Population Dynamics, British Ecological Society Symposia*, **20** (ed. R. M. Anderson, B. R. Turner & L. R. Taylor), Blackwell Scientific Publications, Oxford.

May, R. M. & Hassell, M. P. (1981), The dynamics of multi-parasitoid–host interactions. *The American Naturalist*, **117**, 234–61.

May, R. M., Beddington, J. R., Horwood, J. W. & Shepherd, J. G. (1978), Exploiting natural populations in an uncertain world. *Mathematical Biosciences*, **42**, 219–52.

Mayse, M. A. and Price, P. W. (1978), Seasonal development of Soybean arthropod communities in east central Illinois. *Agro-Ecosystems*, **4**, 387–405.

Mehta, M. L. (1967), *Random Matrices*, Academic Press, New York.

Mellinger, M. V. & McNaughton, S. J. (1975), Structure and function of succesional vascular plant communities in central New York. *Ecological Monographs*, **45**, 161–82.

Milne, H. & Dunnett, G. M. (1972), Standing crop, productivity and trophic relationships in the fauna of the Ythan estuary. *The Estuarine Environment* (ed. R. S. K. Barnes & J. Green), Applied Science Publications, Edinburgh.

Minshall, G. W. (1967), The role of allochthonous detritus in the trophic structure of a woodland springbrook community. *Ecology*, **48**, 139–49.

Morgan, N. C. (1972), Productivity studies at Loch Leven (a shallow nutrient-rich lowland lake). *Productivity Problems of Freshwaters*, (ed. Z. Kazak & A. Hillbricht-Ilkowska), pp. 183–205. Polish Scientific Publishers, Warsaw.

Mori, S. & Yamamoto, G. (1975), Productivity of communities in Japanese inland waters. *Japanese International Biological Programme Synthesis*, Volume 10, University of Tokyo Press, Tokyo.

Moriarty, D. J. W., Darlington, J. E. P. C., Dunn, I. G., Moriarty, C. M. & Tevlin, M. P. (1973), Feeding and grazing in Lake George, Uganda. *Proceedings of the Royal Society*, B, **184**, 299–319.

Morris, R. F. (ed.) (1963), The dynamics of epidemic spruce budworm populations. *Memoirs of the Entomological Society of Canada*, **31**.

Murphy, P. G. (1975), Net primary productivity of tropical, terrestrial ecosystems. *Primary Productivity of the Biosphere* (ed. H. Leith & R. H. Whittaker), Springer-Verlag, New York.

Newton, I. (1979), *Population Ecology of Raptors*, Poyser, Berkhamsted.

Nicholson, A. J. & Bailey, V. A. (1935), The balance of animal populations, part 1. *Proceedings of the Zoological Society of London*, **1935**, 551–98.

Niering, W. A. (1963), Terrestrial ecology of Kaipngamarangi Atoll, Caroline Islands. *Ecological Monographs*, **33**, 131–60.

Odum, E. P. (1962), Relationship between structure and function in the ecosystem. *Japanese Journal of Ecology*, **12**, 108–18.

Odum, E. P. (1963), *Ecology*, Holt, Rinehart & Winston, New York.

Odum, E. P. (1969), The strategy of ecosystem development. *Science*, **164**, 262–70.

Orians, G. H. (1975), Diversity, stability and maturity in natural ecosystems. *Unifying Concepts in Ecology* (ed. W. H. van Dobben & R. H. Lowe-McConnell), Junk, The Hague.

Paine, R. T. (1963), Trophic relationships of eight sympatric gastropods. *Ecology*, **44**, 63–7.

Paine, R. T. (1966), Food web complexity and species diversity. *The American Naturalist*, **100**, 65–75.

Paine, R. T. (1971), A short-term experimental investigation of resource partitioning in a New Zealand intertidal habitat. *Ecology*, **53**, 1096–106.

Paine, R. T. (1979), Disaster, catastrophe and local persistence of the Sea Palm *Postelsia palmaeformis*. *Science*, **205**, 685–7.

Paine, R. T. (1980), Food webs: linkage, interaction strength and community infrastructure. *Journal of Animal Ecology*, **49**, 667–86.

Paine, R. T. & Vadas, R. L. (1969), The effects of grazing of sea-urchins, *Strongylocentrotus* sp. on benthic algal populations. *Limnology and Oceanography*, **14**, 710–9.

Parsons, T. R. & Lebrasseur, R. J. (1970), The availability of food to different trophic levels in marine food chains. *Marine Food Chains* (ed. J. H. Steele), University of California Press, Los Angeles.

Phillipson, J. (1966), *Ecological Energetics*, Edward Arnold, London.

Pimental, D. (1961), Species diversity and insect population outbreaks. *Annals of the Entomological Society of America*, **54**, 76–86.

Pimm, S. L. (1976), Existence metabolism. *The Condor*, **78**, 121–4.

Pimm, S. L. (1979a), Complexity and stability: another look at MacArthur's original hypothesis. *Oikos*, **33**, 351–7.

Pimm, S. L. (1979b), The structure of food webs. *Theoretical Population Biology*, **16**, 144–58.

Pimm, S. L. (1980a), Bounds on food web connectance. *Nature*, **284**, 591.

Pimm, S. L. (1980b), Properties of food webs. *Ecology*, **61**, 219–25.

Pimm, S. L. (1980c), Food web design and the effects of species deletion. *Oikos*, **35**, 139–49.

Pimm, S. L. (1982), Food webs, food chains and return times. *Ecological Communities: Conceptual Issues and the Evidence* (ed. D. R. Strong & D. S. Simberloff), Princeton University Press, Princeton, New Jersey.

Pimm, S. L. & Lawton, J. H. (1977), The number of trophic levels in ecological communities. *Nature*, **268**, 329–31.

Pimm, S. L. & Lawton, J. H. (1978), On feeding on more than one trophic level. *Nature*, **275**, 542–4.

Pimm, S. L. & Lawton, J. H. (1980), Are food webs divided into compartments? *Journal of Animal Ecology*, **49**, 879–98.

Post, W. M. Shugart, H. H. & DeAngelis, D. L. (1978), Stability criteria for multispecies ecological communities. *Oak Ridge National Laboratory, Technical Memoranda*, **6475**.

Raitt, R. J. & Pimm, S. L. (1976), Dynamics of bird communities in the Chihuahuan Desert, New Mexico. *The Condor*, **78**, 427–42.

Readshaw, J. L. (1971), An ecological approach to the control of mites in Australian orchards. *Journal of Australian Institution of Agricultural Sciences*, **37**, 226–36.

Rejmanek, M. & Stary, P. (1979), Connectance in real biotic communities and critical values for stability in model ecosystems. *Nature*, **280**, 311–3.

Roberts, A. (1974), The stability of a feasible random ecosystem. *Nature*, **251**, 607–8.

Roberts, A. & Tregonning, K. (1981), The robustness of natural systems. *Nature*, **288**, 265–6.

Root, R. B. (1973), Organization of a plant–arthropod association in simple and diverse habitats; the fauna of collards, *Brassica oleracea*. *Ecological Monographs*, **43**, 95–124.

Rosenzweig, M. L. (1973), Evolution of the predator isocline. *Evolution*, **27**, 28–94.

Rosenzweig, M. L. & Schaffer, W. M. (1978), Homage to the Red Queen II: coevolutionary response to enrichment of exploitation ecosystems. *Theoretical Population Biology*, **14**, 158–63.

Roughgarden, J. (1979), *Theory of Population Genetics and Evolutionary Ecology: an Introduction*, Macmillan, New York.

Ryther, J. H. (1969), Photosynthesis and fish production in the sea. *Science*, **166**, 72–6.

Saunders, P. T. (1978), Population dynamics and the length of food chains. *Nature*, **272**, 189.

Schaffer, W. M. & Rosenzweig, M. L. (1978), Homage to the Red Queen I: coevolution of predators and their victims. *Theoretical Population Biology*, **14**, 135–57.

Schoener, T. (1974), Resource partitioning in ecological communities. *Science*, **185**, 27–39.

Shugart, H. H. & Hett, J. M. (1973), Succession: similarities of species turnover rates. *Science*, **180**, 1279–381.

Shure, D. J. (1973), Radionuclide tracer analysis of trophic relationships in an old-field ecosystem. *Ecological Monographs*, **43**, 1–19.

Sheldon, J. K. & Rogers, L. E. (1978), Grasshopper food habits in a shrub–steppe community. *Oecologia*, **32**, 85–92.

Slobodkin, L. B. (1961), *Growth and Regulation of Animal Populations*, Holt, Rinehart & Winston, New York.

Southwood, T. R. E. (1978), On the effects of size in determining the diversity of insect faunas. *Diversity of Insect Faunas* (ed. L. A. Mound & N. Waloff), Royal Entomological Society Symposium, Blackwell Scientific Publications, Oxford.

Stiles, F. G. (1978), Possible specialization for hummingbird-hunting in the Tiny Hawk. *Auk*, **95**, 551–3.

Summerhayes, V. S. & Elton, C. S. (1923), Contributions to the ecology of Spitsbergen and Bear Island. *Journal of Ecology*, **11**, 214–86.

Tanner, J. T. (1966), Effects of population density on the growth rates of animal populations. *Ecology*, **47**, 733–45.

Teal, J. M. (1962), Energy flow in the salt marsh ecosystems of Georgia. *Ecology*, **43**, 614–24.

Tilly, L. J. (1968), The structure and dynamics of Cone Spring. *Ecological Monographs*, **38**, 169–97.

Thesiger, W. (1959), *Arabian Sands*, Dutton, New York.

Tregonning, K. & Roberts, A. (1978), Ecosystem-like behaviour of a random interaction model I. *Bulletin of Mathematical Biology*, **40**, 513–24.

Tregonning, K. & Roberts, A. (1979), Complex systems which evolve towards homeostasis. *Nature*, **281**, 563–4.

Tuljapurkar, S. D. & Semura, J. S. (1979), Liapunov functions: geometry and stability. *Journal of Mathematical Biology*, **8**, 25–32.

Tuomi, J. & Niemela, P. (1979), Elioyhteisojen monimuotoisuus ja tasapainoisuus. *Luonnon Tutkija*, **83**, 37–42.

Turelli, M. (1978), A reexamination of stability in randomly versus deterministic environments with comments on the stochastic theory of limiting similarity. *Theoretical Population Biology*, **13**, 244–67.

Valiela, I. (1969), An experimental study of mortality factors of larval *Musca autumnalis*. *Ecological Mongraphs*, **39**, 199–225.

Varley, G. C. (1949), Population changes in German forest pests. *Journal of Animal Ecology*, **18**, 117–22.

Varley, G. C. (1970), The concept of energy flow applied to a woodland community. *Animal Populations in Relation to their Food Resources* (ed. A. Watson), Blackwell Scientific Publications, Oxford.

Vincent, T. L. & Anderson, L. R. (1979), Return times and vulnerability for a food chain model. *Theoretical Population Biology*, **15**, 217–31.

Volterra, V. (1926), Fluctuations in the abundance of species, considered mathematically. *Nature*, **118**, 558–60.

Watt, K. E. F. (1965), Community stability and the strategy of biological control. *The Canadian Entomologist*, **97**, 887–95.

Whittaker, R. H. & Goodman, D. (1978), Classifying species according to their demographic strategy: I, population fluctuations and environmental heterogeneity. *The American Naturalist*, **113**, 185–200.

Wielgolaski, F. E. (1975), *Fennoscandian Tundra Ecosystems*, Springer-Verlag, New York.

Wigner, E. P. (1959), Statistical properties of real symmetric matrices with many dimensions. *Proceedings of the Fourth Canadian Mathematics Congress, Toronto*, pp.174–84.

Wilson, D. S. (1980), *The Natural Selection of Populations and Communites.* Benjamin/Cummings, Menlo Park, California.

Witherby, H. F., Jourdain, F. C. R., Ticehurst, N. F. & Tucker, B. W. (1938), *The Handbook of British Birds*, Witherby, London.

Woodwell, G. M. (1967), Toxic substances and ecological cycles. *Scientific American*, **26**, 24–31.

Wynne-Edwards, V. C. (1962), *Animal Dispersion in Relation to Social Behaviour*, Oliver & Boyd, Edinburgh.

Yodzis, P. (1980), The connectance of real ecosystems. *Nature*, **284**, 544–5.

Yodzis, P. (1981), The stability of real ecosystems. *Nature*, **289**, 674–6.

Zaret, T. M. & Paine, R. T. (1973), Species introduction in a tropical lake. *Science*, **182**, 449–55.

Index